A Theory of the Evolution
of Development

A Theory of the Evolution of Development

by

WALLACE ARTHUR
Department of Biology
Sunderland Polytechnic

A Wiley-Interscience Publication

JOHN WILEY & SONS
Chichester · New York · Brisbane · Toronto · Singapore

Copyright © 1988 by John Wiley & Sons Ltd.

All rights reserved.

No part of this book may be reproduced by any means, or transmitted, or translated into a machine language without the written permission of the publisher.

Library of Congress Cataloging-in-Publication Data:

Arthur, Wallace.
   A theory of the evolution of development.

   Bibliography: p.
   Includes indexes.
   1. Evolution. 2. Morphology. I. Title.
QH371.A765 1988   574.4   88-5628
ISBN 0 471 91974 8

British Library Cataloguing in Publication Data:

Arthur, Wallace, 1952–
   A theory of the evolution of development.
   1. Organisms. Evolution
   I. Title
   575

ISBN 0 471 91974 8

Printed at St. Edmundsbury Press Ltd., Bury St Edmunds, Suffolk.

Contents

Preface vii

1. Theoretical Embryology and the Morphogenetic Tree 1
    1.1 Introduction 1
    1.2 Theoretical embryology 2
    1.3 The morphogenetic tree 6
    1.4 Bringing genes into the picture 13
    1.5 Summary 19

2. Interactions Between the Morphogenetic Tree and Darwinian Selection 20
    2.1 Introduction 20
    2.2 Phase change and von Baer's law 24
    2.3 Structural change and morphological complexity. 29
    2.4 Distortional change and heterochrony 32
    2.5 Summary 32

3. Mechanisms for Major Evolutionary Transitions 34
    3.1 Introduction 34
    3.2 Primary divergence 38
    3.3 Escape to simplicity 39
    3.4 Morphological windows 41
    3.5 Escape from competition 44
    3.6 Summary and conclusion 48

4. Relationships with Other Evolutionary Theories 50
    4.1 Introduction 50
    4.2 Résumé of morphogenetic tree theory 51

4.3 Micro- and macromutational theories 53
4.4 Developmental evolutionary theories 56
4.5 Palaeontological and systematic theories 59
4.6 Prospect 63

5. Applying Morphogenetic Tree Theory 65
5.1 Introduction 65
5.2 General evolutionary trends 66
5.3 Origin of higher taxa 73
5.4 Conclusions 83

References 85

Author index 91

Subject index 93

Preface

As an undergraduate I learned with some fascination of the great groups of organisms which are the product of evolution—molluscs, insects, vertebrates and so on—and of their hierarchical arrangement in the genealogical tree of life. As a research student I studied with equal fascination the mechanics of natural selection, as revealed both by selection experiments in the laboratory and by the correlation between genetic variation and ecological factors in the wild. Then, some years later in an idle moment of reflection, it occurred to me that the connection between these two 'fascinations' was very tenuous. To put it in a more specific way: natural selection as revealed by the studies of the microevolutionist did not seem to be a *sufficient* explanation of the pattern of organic diversity as far as we can characterize it, with its major branches, such as the animal phyla, separated by great gulfs in morphospace, and, on the other hand, its endless variations on small themes, such as the 2000 or so species of *Drosophila*.

This is not to say that natural selection is not *part* of the explanation of morphological evolution: of course it is a part, and a very large one at that. But the grand pattern of the evolution of morphology is a result of the interaction between natural selection and the causal structure of development—the target of selection if you like—and any evolutionary theory attempting to explain the production of organic diversity in terms of only 'half' of this interacting complex is, therefore, insufficient. Such a statement should hardly be controversial, but experience suggests that it is often interpreted as if it were.

The aim of this book is to provide an outline causal structure for development, sufficient at least to begin a study of the interaction between development and selection, and to set out the broad nature of this interaction as I perceive it. The resulting body of theory serves in a way to bridge the gap between two pre-existing bodies of evolutionary thought: the genetic and the morphological. Until now, there seemed little in common between gene-based evolutionary theory, with its emphasis on changes in gene frequency and allele substitutions, and its morphological counterpart, couched in the language of von Baer's law, allometry and heterochrony. The causal structure of development that I propose—the morphogenetic tree—has genes as its input and morphology as its output, and consequently its evolution can equally be seen in terms of relative rates of allele substitution at different loci or in terms of a resultant morphological pattern, such as that described by von Baer's law.

One interesting difference between morphogenetic tree theory and *most* previous approaches to evolution is that the emphasis is on creation rather than destruction. It seems odd to me that, although as evolutionists we are interested in the *production* of particular genomes and morphologies, most approaches to the problem, ranging from the population geneticist's investigation of selection on polygenic variation to the palaeontologist's study of mass extinctions, are based primarily on destructive forces. Of course, if creation, in the form of mutation, is essentially random, and organisms are sufficiently 'plastic' developmentally that anything is possible, then the range of actual morphologies is largely explicable in terms of selection and extinction. That is, creative forces permit everything, while destructive forces determine the actual. One way of looking at this book is as an attack on such a viewpoint. Mutations are not random in space (there are mutational 'hotspots' on chromosomes), in time (they are concentrated in periods of cell division) or in type (transitions are commoner than transversions). And, more relevant to my argument herein, their effects are not randomly distributed in developmental time. I advance the view that this last form of non-randomness is important and needs to be incorporated into any sufficient theory of morphological evolution.

I have written this book so as to make it intelligible to as broad a range of biologists as possible. It should be of equal interest to those with all degrees of biological experience from the 'senior undergraduate' who so often features in prefaces to the 'senior professor' who usually does not (presumably because of his comparative rarity and consequent lack of sales potential!). I hope the book will be of interest to geneticists, developmentalists, taxonomists, palaeontologists and any other group of biologists with a concern for major evolutionary issues.

Morphogenetic tree theory, as presented here, has grown out of an earlier presentation in Chapters 9–13 of *Mechanisms of Morphological Evolution* (Arthur, 1984). It may be helpful to readers already familiar with the earlier book if I point out in advance the main distinguishing features of the updated version of the theory presented here. These are: the introduction of the mutational version of the morphogenetic tree (Chapter 1); the distinguishing of phase change, structural change and distortional change (Chapter 2); a more eclectic approach to the origin of novel body plans (Chapter 3); an examination of the links between morphogenetic tree theory and other evolutionary theories (Chapter 4); and a more concerted attempt to examine the predictions of morphogenetic tree theory in the context of particular case studies (Chapter 5).

Finally, a few words of thanks. This book—or at least the vast majority of it—was looked at in manuscript by Steve Gould, Dick Lewontin and Alec Panchen; and I am indebted to them for their numerous helpful suggestions.

The book has also benefited from discussions I have had with Bill Wimsatt and Nick Rasmussen. Parts of the theory advanced here were presented as seminars at Harvard University and at the Universities of Chicago and Newcastle-upon-Tyne. I am grateful to the various seminar audiences for raising many interesting points, some of which led to clarification of my ideas, and some of which I am still thinking about. Finally, I should express my gratitude to the Royal Society for financial assistance with my visit to the USA, which was instrumental in the development of the book.

WALLACE ARTHUR
1988

Chapter 1

Theoretical Embryology and the Morphogenetic Tree

1.1　Introduction 1
1.2　Theoretical embryology 2
1.3　The morphogenetic tree 6
1.4　Bringing genes into the picture 13
1.5　Summary 19

1.1　INTRODUCTION

The aim of this book is to propose a broad outline for the developmental component that is widely acknowledged to be missing from the 'modern synthesis' of evolutionary theory. In this, the first chapter, I discuss what kind of phenomenon an abstract and general theory of development itself should seek to describe, and I propose such a theory—the morphogenetic tree—expressed first in terms of 'causal links', then in terms of loci, and finally in terms of mutations. The subsequent chapters deal with the evolution of morphogenetic tree systems, and describe the relationship between this view of evolution and diverse others ranging from those of 'heretics' such as Goldschmidt (1940) to the modern synthesis itself. The overall evolutionary theory I propose can be thought of as belonging to what Gould and Lewontin (1979) call the 'weak form' of the continental European approach, which acknowledges the importance of developmental constraints yet does not appeal to mystical internal evolutionary forces and, at the population level, regards natural selection as, to use Darwin's (1859) own words, 'the main but not exclusive means of modification'.

While the theory that I advance is of considerable generality, it is nevertheless restricted in two important ways. These restrictions may perhaps be obvious, but I state them at the outset in order to avoid any misinterpretation of the scope of the theory. The first restriction is that I am proposing a theory of the evolution of *organismic structure* and of the *developmental system* through which that structure arises. What follows therefore has little direct relevance to the evolution of other features of organisms, particularly their *behaviour* or the structure of their *enzymes*. There is thus no clear relationship—either synergistic or antagonistic—between the theory presented herein and a theory that is largely restricted to either of these fields such as, in the latter case, Kimura's (1968, 1983) theory of molecular evolution. Of course, the fact that the *theories* are largely independent does not imply that the processes themselves are independent. For example, evolution of morphogenetic trees as outlined subsequently requires selectively driven changes in enzymes involved in development. Yet these changes may constitute a minority among all the evolutionary changes in these enzymes, and enzymes with a key role in development may be only a small subset of enzymes generally. It is for these reasons that theories of morphological and molecular evolution can remain largely indifferent to each other's fate despite the obvious fact that certain molecules control development.

The second restriction to the theory of morphogenetic tree evolution is a taxonomic one. As will become clear, the concept of the morphogenetic tree itself is an abstraction of the development of a *multicellular* organism; and the emphasis is placed on developmental phenomena, such as pattern formation, which necessarily involve populations of cells, rather than on processes, such as cell differentiation *per se*, which can be manifested within a single cell. Thus what follows has no bearing on the evolution of unicellular or acellular groups. On the other hand, the theory is of general applicability within the context of multicellular taxa. It is true that the examples used are mostly metazoan; but I can see no reason to believe that the development and evolution of other multicellular groups, particularly the metaphyta, should not also be seen in terms of the morphogenetic tree.

What follows, then, in this and subsequent chapters, is a proposal for a major new component of evolutionary (and developmental) theory. As with any new theory, plenty of foreshadowings of it, or at least of parts of it, can be found in previous works. I will mention these individually, throughout the book, at points where they seem to be most relevant. For the moment, I will just make two references to earlier work. First, the ideas presented here are based to some extent on three of my own publications (Arthur 1982b, 1984, 1987a) but incorporate considerable modification and extension of the ideas presented therein. Second, as will gradually become apparent, one of the clearest foreshadowings of morphogenetic tree evolution is ironically due to von Baer, whose major relevant work (1828) was *pre*-evolutionary and who was, in later life, strongly anti-evolutionary (see Gould, 1977b).

1.2 THEORETICAL EMBRYOLOGY

Darwinian selection theory was able to absorb genetics to become neo-Darwinism or the 'modern synthesis', and yet has so far been unable to absorb developmental biology to become more truly synthetic. The reason for this is that an abstract and general theory of heredity has been available for some time (since 1866 or 1900, depending on what is meant by 'available'), whereas there has been no corresponding general theory of development. There is no shortage of observational and experimental work on development, and there is even a smattering of models of various components of the overall developmental process; but these do not add up to a general theory, and they are consequently of restricted interest to the evolutionary biologist.

Before proposing a general theory of development, it is important to consider what question the theory is attempting to answer. I will begin this consideration by noting that the development of any multicellular organism is highly complex and involves many *causal links*. Examples of such links are the switching on of one gene by the product of another, the induction of neural plate by underlying mesoderm in amphibian embryos, and the production of different cell types in response to a concentration gradient of a diffusible morphogen as in the French flag model of Wolpert (1968), to which I will return later.

There are three main questions that may be asked about such causal links, each being the central question at its own level of organization in the overall process. I will now deal with these in turn.

First, at the most down-to-earth level, there is the question of the physicochemical nature of morphogenetic agents. For example, in a particular case, we might ask whether a morphogenetic agent is a molecule or (say) a pulse of electrical activity, whether (if the former) it is a large cell-autonomous molecule or a small diffusible one, and finally what its identity actually is (mRNA, AMP, Ca^{2+}, or whatever). It is possible that, at this level, there is no single entity of comparable generality to DNA in molecular genetics. Even if there is, a detailed knowledge of it is unlikely to be necessary for a theory of development at a higher level of organization/abstraction, just as knowledge of molecular genetics was not necessary for an abstract theory of inheritance (or, for that matter, for the incorporation of genetics into the modern synthesis, which *preceded* the Watson-Crick era). Thus we may dismiss questions about the physicochemical nature of agents involved in the causal links of development as fascinating but not essential to our purpose.

Second, one level of organization above this, we may enquire about the way in which a particular causal link works, regardless of the identities of the agents involved in it. For example, a particular developmental process might depend on a single type of gradient such as the one in the French flag system; a more complex kind of gradient such as the 'double gradient' model of Gierer and Meinhardt (1972; see also Meinhardt and Gierer 1974); or on a system whose operation is not based on a gradient at all (see, for example, Goodwin and Cohen 1969). Of course, certain physicochemical identities at the 'lower level' will preclude certain mechanisms at the higher one: if all the agents involved in a particular causal link are large molecules, then the mechanism cannot be based on a diffusion gradient. But equally, a given physicochemical identity does not proscribe all but a single general type of operation. Thus the two levels of enquiry described above are largely separate issues.

While the distinction between these two levels of approach has been noted by many authors (e.g. Slack 1983, Chapter 7), it has generally not been stressed that there is a third level of approach which is more abstract again (in that it refers to a yet higher level of organization), and whose relationship to 'level 2' (gradients, etc.) is similar to that of 'level 2' to 'level 1' (physicochemical identities). The inter-level relationship is, of course, that each level treats the one below it as a black box. For example, it is possible to examine in detail many of the properties of a gradient system without knowledge of the entity whose concentration forms the gradient. Equally, it is possible to treat *causal links*—for example 'inductions'—as black boxes and enquire what pattern of interconnections of causal links underlies the overall development of a multicellular organism. This point will become clearer if we illustrate it with a particular example, and I will now do this, the example being based on Wolpert's (1968) French flag model mentioned earlier (see also Wolpert, 1969; Crick, 1970). I use this model not because it is inherently more realistic than any other, but rather because it is simple and clear, and enables us to examine the fundamental issue with a minimum of distracting detail. (With regard to the 'reality' of gradients and other types of system, it may well be that there is no single general developmental mechanism at this level comparable to Mendel's 2-factor, segregation-fertilization system; that is, at level 2 as well as level 1, it may be less possible to generalize in development than it is in genetics.)

Figure 1.1 illustrates the French flag system in its full detail as a level 2 phenomenon; and also shows how it may be represented in the more abstract form of level 3 with its components being reduced to black boxes. It is necessary at this point to introduce the idea of a *developmental heterogeneity*. Basically, all developmental systems are heterogeneous in a number of ways, and causal links can be seen as the formation of one sort of heterogeneity from another. In the French flag model, we start with a discrete, 3-component heterogeneity (the line of source/sink/other cells). This gives rise to a continuous, monotonic heterogeneity (the gradient of the diffusible morphogen), which in turn produces a discrete, 3-component heterogeneity (the 'tricolour' arrangement, in two dimensions, of three differentiated cell types).

The fact that we end up with a form of heterogeneity similar to the first is a particularly useful feature of this model because it illustrates the important point that the *transference* of heterogeneity from one form to another in a developmental system (which is what the model describes) is different both from the *initiation* of heterogeneity and from changes in the *number* of heterogeneities over a period of 'developmental time'.

In order to make a level-3 version of the French flag model we need to abandon all of the level-2 detail, treating each heterogeneity as a black box, and merely retaining the sequence of causality. Such an abstraction is seen on the right-hand side of Figure 1.1, where each heterogeneity is represented simply as a circle, and arrows indicate causality. This version of the model retains only one additional bit of information, namely the distinction between *morphogenetic* heterogeneities (solid circles) and *terminal* ones (open circles).

We are now in a position to see the real level-3 nature of the French flag

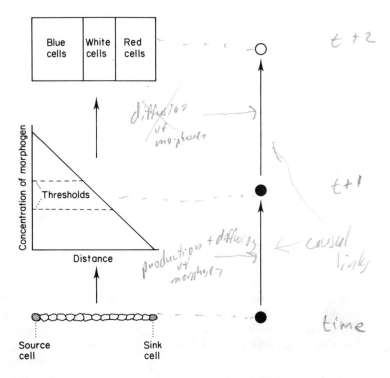

Figure 1.1 The French flag model. Left: version normally given. Right: a more abstract version. ↑ causal link, • morphogenetic heterogeneity, ○ terminal heterogeneity

model. In terms of what *pattern* of *interconnections* of causal links it represents (and recall that this is the central issue at level 3), it is clearly a *linear sequence* developmental system (or what I earlier called a *line* system: Arthur, 1984, Chapter 9).

The question I now wish to pose is: 'what pattern of interconnections characterizes a complete developmental system rather than just a single developmental process?' I have in mind a multicellular organism with a 'simple' life cycle, that is, no metamorphosis, alternation of generations, or other such complexities. These will be dealt with later. For the moment, it is wise to ignore them and to consider a kind of organism, such as a mammal, in which there is only one major developmental phase in the life cycle.

Given such an organism, what pattern of interconnections of causal links characterizes its development? The simplest possibility, of course, is that there is again a linear sequence of links: just a longer one than that typifying a component of development such as the French flag system. It is surprising how frequently in the developmental literature statements occur that seem to suggest this, by the use of words such as sequence, series, or chain. For example,

Shumway (1932) puts it as follows: 'Every stage, or element of a stage, forms a link in a continuous *chain* of cause and effect' (my italics). I suspect not only that this is wrong, but also that those making such statements would admit that they are wrong. Probably, this is just a comparable loose usage to that of an ecologist who talks about a food chain when in fact he knows it is really a web.

The simplest possible answer to our question is that the pattern of causal interconnections underlying development is a linear sequence. At the other extreme, there is the possibility that the pattern is so complex and irregular that we cannot hope to describe it and/or that it is so different in different organisms that we cannot hope to generalize about it. I do not believe this any more than I do the idea of a 'linear sequence development', but if this disbelief is misguided then so is the whole approach adopted here.

In between the 'linear sequence' and 'undecipherably complex' possibilities lie many other possible patterns of interconnection, and it would be futile to attempt to describe all of them. Rather, I will, in the following section, describe the one pattern that I am proposing—a hierarchical or tree-like pattern—and will discuss the reasons for making this proposal as well as the extent to which it is in agreement with the views of others who have given consideration to level-3 developmental phenomena.

Before ending the present section, I should make two further points. First, my reason for concentrating on level-3 phenomena is partly a conviction that it is possible to achieve a wider generalization at this level of development than at lower ones, and partly a conviction that this level is most relevant to evolutionary theory. It is for the reader to decide whether he agrees with these convictions. Second, it must be stressed that level-3 phenomena really can be considered in isolation from those at level 2. Having reached the level of abstraction appearing on the right-hand side of Figure 1.1, it no longer matters that this was derived from a 'single gradient with localized source and sink' system; our circles and arrows could equally have been obtained from any level-2 system, including those not based on gradients at all.

Finally, as an aside, I should say that I chose the term 'theoretical embryology' deliberately as the title for this section, despite the aura of descriptiveness that sometimes accompanies the term embryology (in contrast to the experimental overtones of 'developmental biology'). There were two reasons for this decision. First, 'theoretical developmental biology' is a cumbersome label, and we ought not to dismiss aesthetic considerations when devising phrases. Second, it is literally true that the morphogenetic tree is concerned with *embyros*. It may well not be a useful idea in relation to other developmental processes, such as post-embryonic growth, where other general concepts, such as allometry, become more appropriate.

1.3 THE MORPHOGENETIC TREE

The idea that the pattern of interconnection of causal links underlying the overall development of an organism such as a mammal is a hierarchy can be

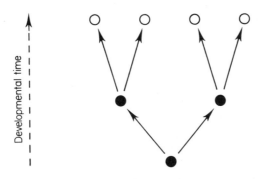

Figure 1.2 The morphogenetic tree ('simple' version). ↑ causal link, • morphogenetic heterogeneity, ○ terminal heterogeneity

expressed in a very simple diagrammatic way: see Figure 1.2. This clearly indicates that as well as *transference* of heterogeneity from one form to another through developmental time, there is an *increase* in the *number* of simultaneously existing heterogeneities over time. This is achieved by each one heterogeneity at any temporal stage in development t_i causing the production of more than one heterogeneity (two in the special case considered in Figure 1.2) at the subsequent stage t_{i+1}.

It is not difficult to envisage a mechanism through which this could happen, and the precise way in which it happens need not concern us as it is a level-2 phenomenon. However, a hypothetical example might help to reinforce the picture I am trying to build up. Suppose, in the context of the French flag model, that not only do cells respond to the thresholds shown in the concentration of the diffusible morphogen by differentiating into 'red', 'white', and 'blue' cells, but also that, regardless of their fate in the 'colour' dimension, they respond to a threshold in the centre of the gradient (not shown in Figure 1.1), those cells to the left-of-centre becoming specified as anterior compartment cells and those to the right as posterior. In this case, one heterogeneity (the gradient) has produced two.

Having described what is meant by a morphogenetic tree, namely a hierarchical pattern of interconnection of causal links in development, a few words should be said about what led me to favour this particular pattern out of a wide range of possible patterns. In fact, a very elementary observation on the development of organisms with 'simple' life cycles suggests, but of course does not prove, that the underlying causal structure of development is hierarchical. Basically, causal links, whether described in the embryological literature as inductions, cytoplasmic localizations, gradients or whatever, operate in relatively restricted spatial areas. These areas, often referred to as embryonic fields—or simply fields—are generally thought to have an upper size limit of about 1 mm (see, for example, Wolpert, 1971). During early cleavage, the whole organism constitutes a single field. As it gets bigger and more complex, it must subdivide into an increasingly large number of different fields, both because of the straightforward effect of size, and because the occurrence of different developmental processes in different places clearly shows that the causal links operating in those places are in some way different from each other. Since the layout and nature of fields at t_{i+1} is dependent on the smaller number of fields at t_i, the causal structure of the overall process is hierarchical.

I turn now to foreshadowings of the morphogenetic tree concept in the work of other authors. Indeed, foreshadowing is in some cases too weak a word: 'statement' might be better, because the proposal made at some length above (and in Arthur, 1984, Chapter 9) has been made quite explicitly by at least two authors, and probably many more. However, a cautionary comment is necessary at this point. Many authors refer to development as being, in some unspecified way, hierarchical. For example, Koestler (1970) has said: 'Morphogenesis proceeds in an unmistakably hierarchical fashion'. This is fine, as far as it goes, but since development is hierarchical in an obvious way—it is a hierarchical cell 'lineage'—it is difficult to know what such a very general appeal to hierarchies means. What we must look for specifically are claims that the *causal structure* of development is hierarchical.

Two explicit claims of this sort are those of Slack (1983) and Sang (1984). In the first chapter of his book, Slack makes the following statement (p.5): 'It is very important to emphasize that even the basic body plan is not specified all at once but is formed as a result of a *hierarchy of developmental decisions*' (my italics). Shortly afterwards (p.6) he refers to 'the conviction that the basic body plan arises from a *hierarchy* of decisions between determined states' (Slack's italics!). In the rest of the book, similar statements appear intermittently. Sang (1984, p.329) states (in conjunction with a reference to the Britten-Davidson model of gene regulation: see Britten and Davidson 1969, Davidson and Britten 1979), 'morphogenetic complexity depends on the activation/inactivation of gene *sets*, not single genes; possibly in a *hierarchical sequence of determinative steps*' (the second italicization is mine). These statements are particularly interesting, both because of the clear reference to hierarchy of causality and because of the tentative way in which both authors put them. Slack's use of 'conviction' and Sang's 'possibly' indicate clearly that both authors regard hierarchical causality as a possibility rather than a certainty, thereby acknowledging the existence of alternative causal structures, of which, as we noted earlier, there are many.

While the statements of Slack (1983) and Sang (1984) are particularly clear, and obviously closely related to the morphogenetic tree concept, many earlier authors have also advanced the idea of developmental hierarchy. Waddington's (1957; p.29) 'epigenetic landscape' incorporates a hierarchical pattern of developmental decisions. The idea of developmental hierarchy is central in the work of Løvtrup (1974) and Riedl (1978). As we will see in the following section, there is even a statement that relates quite closely to the mutational version of the morphogenetic tree concept in the first edition of *The Origin of Species*.

So far in this section I have described in outline the concept of the morphogenetic tree, and have indicated that the idea is not a new one, though I take it considerably further, in this book, than earlier authors have done. It is now necessary to examine the simplified version of the tree given in Figure 1.2, and to expose, and subsequently *dis*pose of, certain unrealistic features. Having done this, I will sketch two more complex pictures of the tree, which should be more readily acceptable.

The obvious way to start this formulation of a more realistic picture is to list the undesirable simplifications inherent in the 'simple' tree of Figure 1.2. There are basically eight of these, but I list just seven below. The eighth merits more extensive discussion and will be dealt with later.

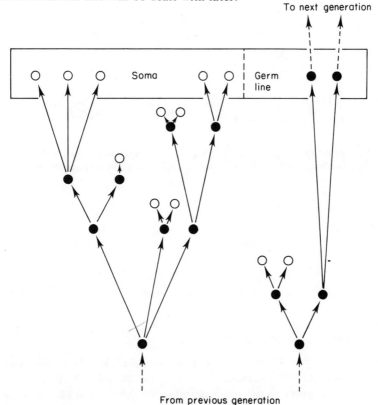

Figure 1.3 The morphogenetic tree ('complex' version). 'Box' indicates adult; inter-generation causal links shown as dashed arrows

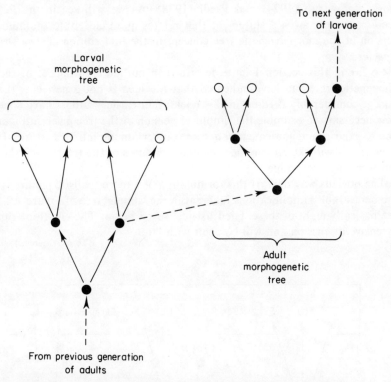

Figure 1.4 Double-tree system underlying the development of a holometabolus insect. Inter-generation and inter-lifestage causal links shown as dashed arrows

1. The developmental process begins with just a *single* heterogeneity.
2. That heterogeneity arises *de novo* at the beginning of the life cycle.
3. All non-adult heterogeneities are morphogenetic.
4. All adult heterogeneities are non-morphogenetic.
5. The increase in the number of heterogeneities is through bifurcation only.
6. There are definite developmental stages, i.e. the heterogeneities in different 'branches' are synchronous in developmental time.
7. There is only one tree per life cycle.

Almost certainly, 1–6 are not true of *any* multicellular organisms, while 7 is not true for *some* of them (those with 'complex' life cycles). A more complex version of the morphogenetic tree that takes account of points 1–6 is given in Figure 1.3. A double tree representing the development of a holometabolous insect such as *Drosophila*, where the adult begins its development as a series of imaginal discs and histoblasts within the larva, is shown in Figure 1.4. Figure 1.4 is based on two simple trees rather than two complex ones, because otherwise the diagram would become inscrutably complicated and because, at any rate, the extra realism of Figure 1.3 is unnecessary in relation to the point that Figure 1.4 is making.

Both of Figures 1.3 and 1.4 merit some further discussion, and I will take them in turn. In Figure 1.3, it is the connection between generations that is most deserving of our attention, which relates to points 1 and 2 in the above list. A developmental heterogeneity may originate in two ways, one of which involves its production *de novo* from a previously existing homogeneous state. Although such a process is known to be possible in theory (Turing, 1952), and although it may actually operate in practice at certain points in the development of some organisms (see, for example, Maynard Smith and Sondhi, 1961), the version of the morphogenetic tree given in Figure 1.3 represents a denial that the first heterogeneities in a particular life cycle originate in this way. Rather, this version of the tree involves a heterogeneity 'carry-over' from one generation to the next; that is, it involves not only the majority of heterogeneity transfers *within* a generation, but also those occurring beteen generations, originating from previously existing heterogeneities. Admittedly, those heterogeneities crossing what I have called the 'generation bridge' (Arthur, 1987a) exist in a highly condensed and codified form, but they are there nonetheless. Thus in morphogenetic tree theory the concept of *initiation* of heterogeneity is not central, basically because such an event is relegated to an unimportant role in developing systems, and is certainly *not* seen as a key event at the beginning of a life cycle.

The question of how many heterogeneities are carried over from one generation to the next is of course unanswerable at present, and may be highly variable among different kinds of organism. The fact that Figure 1.3 depicts two such heterogeneities is of no consequence: diagrams of this sort are always special cases as well as general pictures, and can easily generate confusion because of this. But what *is* important is that the number of developmental heterogeneities crossing the generation bridge will be conserved from one generation to the next, at least over a short span of evolutionary time.

Turning to the issue of complex life cycles, and thus to Figure 1.4, the only point that needs to be made is that this picture of complex life cycles as double (and in the general case multiple) morphogenetic trees can be extended to deal with all sorts of other complex developmental systems in addition to the case of holometabolous insects pictured in the Figure. All that is required are differences of detail; and sometimes it would be necessary to represent the system as a 'double complex tree' rather than a 'double simple' one. For example, in the production of a sporophyte fern from a gametophyte, it would be necessary to draw the connection to the base of the sporophyte's morphogenetic tree from a morphogenetic heterogeneity in the mature 'adult' gametophyte, and so it would be necessary to have a soma/germ-line distinction, as there is in Figure 1.3.

The vital point about complex life cycles generally is that in morphogenetic tree theory they are seen as a pair (or larger number) of linked causal hierarchies. This is important, because I have heard an alternative view advanced, namely that complex life cycles argue against the idea of hierarchical development altogether.

We now come to the eighth deficiency in the original picture of the morphogenetic tree (Figure 1.2), which, unlike the first seven, is also embodied in the tree-systems of Figures 1.3 and 1.4. In order to appreciate the nature of this deficiency, we need, in a sense, to go back to basics.

Often, in developmental biology, a distinction is recognized between the three processes of cell differentiation, pattern formation and morphogenesis (see, for example, Garrod, 1973). Some authors, including Thom (1983) and myself (Arthur, 1984), have argued that the last two of these should be treated together as the among-cell component of development. Other authors (Slack, 1983) use the term regional specification for, essentially, this among-cell component. In both/all of the above processes, Wolpert's (1969) concept of positional information is relevant, because the diverging developmental fates of cells/tissues/regions may be controlled by information of this kind, perhaps operating through a gradient-type system. However, *all* of these terms and concepts relate to one 'side' of the developmental process, that of 'becoming different'. There is another, in a sense complementary, side to development, namely 'staying the same'. Three phenomena, which are often not clearly distinguished from each other, fall under this heading.

1. *Canalization*. This is the term used for the tendency of a developmental process to produce the same result despite perturbations of a variety of kinds, both genetic (e.g. presence of a defective recessive gene in heterozygous condition) and environmental (e.g. experience of unusual temperatures during embryogenesis). The concept of canalization is closely associated, of course, with the name of C.H.Waddington (1942, 1957).
2. *Co-ordination*. During the normal development of an individual who does not experience any such perturbations, there is still a 'staying the same' phenomenon, which I call co-ordination. This is the ability of different parts of the developing system to proceed in the correct manner and time-sequence in relation to each other. An obvious example is the developmental simultaneity and mirror-image likeness of the complementary halves of a bilaterally symmetrical body plan.
3. *Repeatability*. This is the label I give to the tendency of overall embryogenesis to produce near-identical results in different individuals of the same species.

The main inadequacy of the morphogenetic tree concept is that it does not explain these 'staying the same' aspects of development. This point can be made more clearly if we focus, in particular, on co-ordination. Different developmental processes occurring in the same organism have a considerable potential, in theory at least, to get out-of-phase with each other. This potential is somewhat analogous to the increasing divergence between the actual and intended trajectories of a missile, given a very slight initial error in the angle of firing. Since developmental systems do tend to remain highly co-ordinated throughout their progress from egg to adult, it seems reasonable to suppose that there is some mechanism for keeping different subsystems in line with each other. Even in the 'complex' morphogenetic tree of Figure 1.3, there is no such mechanism, because different 'branches' of the tree remain separate from their point of divergence right up to the adult stage.

At present, the only suggestion that it seems reasonable to make is the very general one that, superimposed on the basic structure of the morphogenetic tree, as depicted in Figure 1.3, there is some sort of cross-linking of the heterogeneities in different 'branches'. Interestingly, Waddington makes a similar proposal (with regard to explaining canalization). He states (1957, p.23): 'canalizations are more likely to appear when there are many cross-links between the various processes'.

While I make no bones about the fact that I see the lack of a *clear* picture of co-ordination/canalization as the biggest single inadequacy of the morphogenetic tree concept, there are nevertheless two positive points that I should make at this stage. First, I have a conviction (and it is only that) that when the 'staying the same' side of development can be more clearly pictured at organization level 3, it will add to, rather than replace, the morphogenetic tree, that is, it will not cause a revision of the idea that the causal structure of the 'becoming different' side of development is hierarchical. Second, although the idea of developmental constraint is often connected with that of canalization, the morphogenetic tree *does* incorporate developmental constraint. Basically, the degree of constraint on any developmental heterogeneity in the tree structure can be measured in terms of the number of subsequent heterogeneities that depend upon it, or, equivalent to this, on how far 'down' the tree it occurs. (There is a close parallel here with Wimsatt's (1986) concept of 'generative entrenchment' in his 'developmental lock' model; see also Schank and Wimsatt, 1986.) Indeed, this kind of constraint may have a key role in determining the evolution of morphogenetic trees, as we will see in the subsequent chapter.

1.4 BRINGING GENES INTO THE PICTURE

So far, the morphogenetic tree has been described only in terms of causal links and developmental heterogeneities, and there are two problems inherent in this kind of description. First—and probably less important—while it is obvious that such links and heterogeneities exist, they are difficult to delineate from each other, and consequently, in a sense, difficult to define. Second, these developmental 'units' cannot readily be connected with *evolutionary* theory. In order to make a clear connection between the morphogenetic tree itself as a developmental concept and morphogenetic tree evolution, we need to use a kind of unit of description that is meaningful in both contexts. This immediately brings us to the gene, and the need to describe the morphogenetic tree in genetic terms.

Of course, 'gene' itself is ambiguous. Molecular biologists frequently use gene and mutation to mean locus and allele, respectively, thus gene=locus.

Yet when the population geneticist uses the term gene frequency, he means allele frequency, and so gene=allele. I point this out not because of a liking for definitional hair-splitting but rather because the distinction between the two usages of 'gene' becomes crucial in the argument to be developed, as will be seen shortly.

I will now proceed to describe the morphogenetic tree first in terms of loci and then in terms of mutations (and hence alleles). As will become clear, the latter version is both more realistic and more useful for evolutionary theorizing. However, there is a good reason to reach the mutation-based version indirectly, in two stages. The reason is that if we proceed from heterogeneities to loci to mutations, we add assumptions to the theory bit by bit, which reveals these assumptions more clearly than if they were added all at once.

Before starting the locus-based version of the morphogenetic tree, I should stress that the loci concerned are only those with some role in the developmental process. I have earlier called these D-loci (Arthur 1984, Chapter 10), and made it clear that this category is heterogeneous at the molecular level in that it probably includes loci that make proteins with a developmental function as well as those that more directly control the activities of other loci (e.g. via RNA) without making a protein product. As before, we need not be concerned with this level of detail: it does not matter for the purposes of the theory how particular D-loci work. What does matter is the number of these loci, their magnitude of phenotypic effect, and the distribution of their activities through developmental time; and we will now proceed to concentrate on these three things. One final point here: I will simply refer to 'loci' from here on. There is no need to keep saying D-loci, as I will not be discussing any loci with no developmental role; all loci referred to are therefore D-loci by implication in the following discussion.

Only one assumption is necessary to 'translate' the morphogenetic tree concept from heterogeneities to loci. This assumption takes the following form: whatever the correspondence between number of loci and number of heterogeneities/causal links, there is no systematic trend in this correspondence through developmental time. The actual nature of the correspondence is not important. Whether there is a precise and constant 1:1 correspondence (which seems most unlikely) or a more complex and more variable correspondence need not concern us providing that the variability is not expressed as a time trend.

Making just this single additional assumption, we can express the morphogenetic tree as a hierarchical pattern of causal interaction among the loci governing development—a pattern for which there is some recent evidence (see e.g. Ingham and Martinez-Arias, 1986). This 'translation' requires no separate visual presentation because the pictures of the morphogenetic tree provided earlier (e.g. Figure 1.2) can simply be interpreted as representing loci rather than heterogeneities. However, I now wish to *extend* the tree concept by bringing in an extra dimension: magnitude of phenotypic effect. If we make the further assumption that the earlier a locus begins to affect the developmental process the greater its impact on that process, and consequently on the adult pheno-

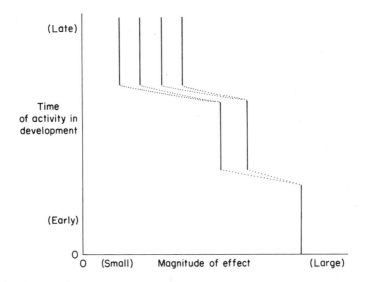

Figure 1.5 Locus-based version of morphogenetic tree. Each vertical bar indicates the activity of a locus. Causal links are shown here as dotted lines. In the genetic context, a causal link is the switching on/off of one gene (directly or indirectly) by the activity of another

type, then we end up with an enlarged picture, as shown in Figure 1.5. This incorporates the original idea of a hierarchy, as expressed in Figure 1.2, but also the idea of diminishing effect with lateness of onset of action. (Note that the *duration* of action is shown as being constant across different loci, which is doubtless a simplification.)

Although this picture of the morphogenetic tree allows for some variation in the magnitude of effect of different loci acting at the same developmental stage, it should be emphasized that each *individual locus* is represented as having a single, specified magnitude of effect. It is now pertinent to ask the related questions of (a) how we measure this, and (b) whether it is acceptable to picture a locus in this way.

In fact, it is unreasonable to picture loci like this, because we measure the magnitude of effect of a locus by examining mutations of it, and it is clear that all loci that have been extensively studied exhibit variation in the magnitude of phenotypic effect of their various mutations. Thus a locus should be pictured as having not a single, hypothetical, magnitude of effect, but rather a *distribution* of effects characterized by a mean and variance, and ideally built up from an exhaustive study of all known mutations of the locus concerned. If we treat the bars representing magnitude of effect in Figure 1.5 as means, and replace them with the distributions to which they relate, then we get the picture shown in Figure 1.6. This is the mutation-based version of the morphogenetic tree. (In fact, each *mutation's* magnitude of effect is not fixed, as it depends to some

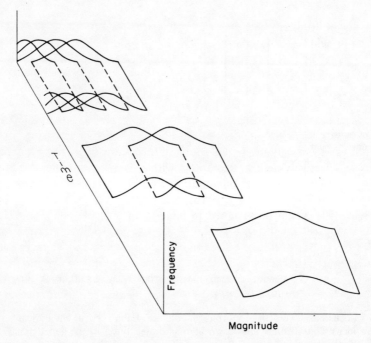

Figure 1.6 Mutation-based version of the morphogenetic tree. The tree is now three-dimensional. The third dimension, projecting upwards from the time/magnitude plane, is the frequency of occurrence of mutations with different magnitudes of effect at each of the loci shown in Figure 1.5. Causal links are not shown here for simplicity (the diagram is too complex already!)

extent on the genetic background and the environment, but such complications need not concern us here.)

Three attributes of the distributions shown in Figure 1.6 should be noted. First, and least important, they are drawn as normal curves. This is entirely arbitrary, and undoubtedly the distributions in reality are of a more complex shape. (One feature to be anticipated, perhaps, is a disproportionate increase in variance with increasing mean.) Second, if we move backwards through development, the mid-points of the distributions move from left to right. There is nothing new in this: the same phenomenon was observed in the locus-based version of the tree (Figure 1.5). Third, at some point in our journey backwards through development, *the left-hand tails of at least some of the curves detach themselves from the time-axis*. This leaves a gap indicating that mutations of small effect—those in between zero on the magnitude axis and the start of the left-hand tail—are not possible. Moreover, *this phenomenon becomes more pronounced* the further 'back' we go towards the beginning of development. The rationale behind this proposal is as follows. The genes concerned are deemed to be making developmental *decisions*. For example, they determine whether a particular region will be anterior or posterior, or whether it will contribute to one body segment or another; or they determine whether a given group of cells will differentiate into one tissue type or another. Mutations in these genes either do or do not affect their decisions. If not, then from a developmental point of view, the mutation concerned has a zero effect. If so, then the magnitude of effect is largely determined by the magnitude of the decision.

The proposal that is being made, then, is that at least some of the loci governing key decisions in early development are not capable of undergoing effective mutation without producing a major effect, albeit their mutations will be somewhat varied in magnitude. Of course, *all* loci are likely to be capable of undergoing selectively-neutral mutations with a zero magnitude of effect, such as might occur in the case of mutation in a redundant third base in a gene making a morphogenetic enzyme (but see Clarke, 1970); so all distributions in Figure 1.6 should be assumed to include an invisible element sitting on the time-axis. The important assertion is not that zero-effect mutations are possible at early-acting loci, but rather that there is a gap, on the magnitude scale, between such mutations and the rest of the distribution appropriate to the locus concerned. If this is true, it means that in evolution it is impossible to modify some early developmental processes by gradual accumulation of micromutations. This is where morphogenetic tree theory parts company with strict neo-Darwinism, but I leave consideration of such evolutionary issues to the following chapters.

At this point, it is appropriate to enquire about the relationship between the genetic versions of the morphogenetic tree (particularly the mutational one) and the views of other geneticists, just as I indicated earlier the relationship between the original version of the morphogenetic tree and the ideas of several developmental biologists (who are in some cases the same people wearing different hats!).

In fact, the mutational version of the tree is a compromise between two extreme genetic viewpoints. One is the view, attributed to Mather (1941, 1943a,b,c), that there are different kinds of *loci* each typified by a given magnitude of phenotypic effect. (Specifically, Mather distinguished the small-effect 'polygenes' from large-effect major genes or 'oligogenes'.) The emphasis, in this view, is on *among-locus* variation in the magnitude of effect. An example of the counter-view to this can be found in Fincham (1983), who states that: 'This concept of polygenes as distinct from major genes has been largely discredited'. Fincham adds: 'The same genes which show this insignificant, or barely significant, variation can also, through other kinds of mutation, give rise to alleles with extreme or even lethal effects on the phenotype. Whether a gene is associated with a large or a small effect depends on the variants that are present in the population under study.' In this view, it is clearly among-allele, *within-locus* variation in magnitude of effect that is emphasized, the degree of effect thus being seen as a property of a particular mutation, not a particular locus.

The mutational version of the morphogenetic tree is a compromise between these two views in that it explicitly recognizes both within-locus *and* among-locus aspects of 'magnitude of phenotypic effect'. Moreover, it recognizes within and among developmental stage components of the among-locus variation, and assigns a time-trend to the latter.

While the recognition of a pronounced among-locus, and particularly among-stage component to the variation in magnitude of effect may run contrary to currently popular genetic views (as illustrated by the quotations from Fincham, 1983), it does have a long history. Consider, for example, the following comment made by Darwin (1859, Chapter 13), in which he is essentially using 'monstrosity' for large-effect mutation and 'slight variations' for small-effect, polygenic mutations. 'It is commonly assumed, perhaps from monstrosities often affecting the embryo at a very early period, that slight variations necessarily appear at an equally early period. But we have little evidence on this head: indeed the evidence rather points the other way; for it is notorious that the breeders of cattle, horses, and various fancy animals, cannot positively tell, until some time after the animal has been born, what its merits or form will ultimately turn out.' Admittedly Darwin goes on to muddy the issue by launching into one of his occasional Lamarckian interludes but, nevertheless, the idea of a connection between developmental stage and magnitude of phenotypic effect of mutations is clearly stated.

The connection I have made so far between the concept of the morphogenetic tree and some comments of Darwin, Mather and Fincham relates entirely to the new dimension—magnitude of effect—that was added to the tree concept in its genetic guise. However, the genetic version still also incorporates the tree itself, that is, the causal hierarchy. It is clear that Waddington (1940) had something like the morphogenetic tree in mind when he was discussing what he called the 'branching-track system' of developmental decisions. It is perhaps unfortunate that he immediately added a third ('vertical') dimension to this, turning it into the 'epigenetic landscape'. The motivation behind this—a desire to incorporate canalization—was understandable, but the result was a rather unwieldy idea which has tended to remain as a vague picture rather than becoming a basis for productive evolutionary theorizing. Indeed, Waddington himself never adequately explained how he saw the connection between his epigenetic landscape concept and neo-Darwinian evolutionary theory, despite the fact that he went on to develop other ideas that were clearly selection-based, such as genetic assimilation.

Two final points should be made about the genetic versions of the morphogenetic tree. First, many genetic studies have revealed the passage of pattern-forming substances across the 'generation bridge', and thereby revealed the reality of the idea that the first heterogeneities in development arise from parental (maternal) heterogeneities rather than arising *de novo*. One good example is the maternally inherited substance determining dextrality in the snail *Lymnaea peregra* (see Freeman and Lundelius 1982; and also the earlier work of Boycott and Diver, 1923; Sturtevant, 1923; Boycott *et al.*, 1930). Another is the product of the *esc* gene in *Drosophila melanogaster* (Struhl, 1981), which also exhibits maternal inheritance. Many other examples could be given.

The second point is that there may, in a sense, be two interacting genetic hierarchies in development. It is possible that, within a cell, the pattern of gene-switching causing differentiation into a particular cell-type is hierarchical (see the model of Britten and Davidson, 1969, mentioned earlier); and that hierarchies of this sort are activated by causal links towards the tail-end of the overall among-cell hierarchy of pattern formation/regional specification. However, I will not develop this idea further. By now we have a sufficient number of proposed generalities about development to use as a basis for a theory of long-term morphological evolution, and, after briefly summarizing these generalizations, I will now turn from development itself to its evolution.

1.5 SUMMARY

The main logical sequence of this chapter, reduced to its bare essentials, is as follows:

1. The modern synthesis lacks a coherent developmental component.
2. The main reason for this is that there is no general theory of development comparable to Mendel's theory of inheritance.
3. The three levels at which the problems of development may be approached are noted, and the central question of each is stated.
4. It is asserted that the highest of the three levels is the most likely to produce generalizations about development itself, and also the most relevant to evolutionary theory.
5. The central question at this level is: what is the pattern of interconnection between the various causal links that make up the overall developmental process of a multicellular organism? (The nature of each link is treated as a black box at this level.)
6. It is proposed that, of the many possible patterns, the true one is basically hierarchical. Hierarchical causal structure of development is labelled the morphogenetic tree. This is a potentially general theory of development at the level described.
7. The morphogenetic tree is then described in genetic terms, and a new dimension is introduced: magnitude of phenotypic effect.
8. The 'ultimate' version of the tree concept—the mutational version—depicts a hierarchical genetic contribution to development, with few early-acting genes prone to large-effect mutations, and many late-acting ones which tend to have small-effect mutations.
9. Genes contributing to the morphogenetic tree are more developmentally constrained the earlier they act and the more numerous the later developmental processes that depend on them.
10. This will give rise to a form of evolution that is incompatible with strict neo-Darwinism but is not incompatible with a more liberal neo-Darwinian view; and indeed it is stressed here and in the following chapters that morphogenetic tree evolution is best seen as an extension of the modern synthesis, and not as an attempt to replace it.

Chapter 2

Interactions Between the Morphogenetic Tree and Darwinian Selection

2.1 Introduction 20
2.2 Phase change and von Baer's law 24
2.3 Structural change and morphological complexity 29
2.4 Distortional change and heterochrony 32
2.5 Summary 32

2.1 INTRODUCTION

In the previous chapter, I formulated the hypothesis that the causal structure underlying development is basically hierarchical, that is, it takes the form of a 'morphogenetic tree', which can be expressed in terms of either causal links or genes. In the present chapter, I examine the evolution of morphogenetic tree systems under the influence of natural selection. I use 'Darwinian' in the title to indicate that there is nothing new in the kind of selection envisaged, and in particular, I leave consideration of 'n selection' (Arthur, 1984) to Chapter 3. Moreover, I consider only the simplest kind of Darwinian selection: directional. Stochastic processes, as well as more complex forms of selection, such as balancing, are ignored. This is not because I suppose these to be generally unimportant in evolution, but rather because the emphasis herein is on long-term evolutionary *change* as opposed to maintenance of variation, and on *adaptive* changes as opposed to those that are selectively neutral.

While this chapter does not set out to say anything new about selection, or indeed about the developmental system on which selection acts (which was characterized in Chapter 1), it most certainly *does* attempt to say something new about the interaction between the two—indeed this interaction provides the focus of attention. The difference between the theory to be developed below and strict neo-Darwinian theory is that evolutionary patterns at the morphological level are seen herein as products of the interaction between natural selection and a definite internal structure of development—the morphogenetic tree—rather than a product of selection alone. The sometimes vague idea of developmental constraint is made explicit and is seen as increasing in severity with distance 'down' the morphogenetic tree (i.e. towards the beginning of development). Morphology is viewed as being moulded, in *evolutionary* time, by the opposition between selection, on the one hand, and a degree of developmental constraint that varies systematically in *developmental* time, on the other. Because of this emphasis on the interaction between an external, ecological force and an internal causal structure incorporating the idea of developmental constraint, the theory of morphogenetic tree evolution is unashamedly fence-sitting in relation to Gould's (1977a) second 'eternal metaphor', that is, the question of whether evolution is largely directed by forces external or internal to the organism. If the theory developed herein is basically correct, then not only are theories based entirely on internal forces fundamentally flawed (as is now generally accepted anyway), but so are theories based entirely on external forces. Theories of the latter kind view the organism as something internally amorphous that can readily be moulded by selection so as to render its design optimal for any given environment. (This is the 'adaptationist programme', as criticized by Gould and Lewontin, 1979.)

The approach I adopt here is to take the morphogenetic tree structure described in the preceding chapter for granted, to enquire about the modes of evolutionary change open to such a structure, and to relate these to already known patterns in morphological evolution, particularly von Baer's law, the trend towards increased morphological complexity, and heterochrony.

What modes of evolution, then, are possible for a morphogenetic tree system? There are basically three: I have termed them (1) phase change, (2) structural change and (3) distortional change (Arthur, 1987a), and will now briefly outline what is meant by these terms before devoting a section to each.

Each causal link in the morphogenetic tree can be thought of as embodying an instruction (or set of instructions) that is being transmitted from one developmental process/stage to another. If a mutational change occurs such that the nature of the instruction (or signal or message—call it what you will) changes, but its relation to the other causal links making up the morphogenetic tree remains unchanged, then we have what I call a *phase change*. It should be noted that phase changes make organisms different from each other morphologically, but do not make one organism more complex than another. An example of a phase change is the switch from a dextral to a sinistral gastropod, or vice versa. Here, an instruction to the early cleavage system to be asymmetrical in a particular way is replaced by an instruction to be asymmetrical in precisely the opposite way. Admittedly, the two instructions may differ in that one is an 'active' message, the other a sort of 'default' message (Freeman and Lundelius, 1982); also, the switch from one message to the other may not be entirely without consequence for later causal links (Gould *et al.*, 1985). Nevertheless, the sinistrality/dextrality switch is an example of a phase change rather than of either of the other two types of change described below.

While the number of causal links comprising the morphogenetic tree of one of a pair of sibling species must be more-or-less identical to the number in the other member of the pair, the number of links in the morphogenetic tree of a complex metazoan such as an insect or vertebrate must be much greater than the corresponding number in a simpler organism such as a sponge, coelenterate or flatworm. Thus it is clear that, while in the short term the structure of the morphogenetic tree may be relatively constant, and morphological evolution proceeds largely by phase changes (usually of much smaller effect than the chirality mutation), in the long term the very structure of the tree must alter. This *structural change* can involve addition of previously nonexistent links, deletion of previously existing ones, or both. Structural changes *do* make one organism more or less developmentally and morphologically complex than another, and while changes in both directions have occurred in evolution, a special challenge to the evolutionary biologist is provided by structural changes that *increase* complexity, as will be discussed in Section 3. (Note that the deletion of causal links can involve deletion of inductive stimulus and/or response. The evolutionary loss of teeth in birds would appear to be an example of loss of stimulus: see Kollar and Fisher, 1980.)

The third and final kind of change to which morphogenetic trees may be subjected is *distortional change*. Here, no causal links are added or deleted, and there is no fundamental alteration in the message embodied in any one of them (such as would occur in a phase change), but rather, the timing of the links in one or more 'branches' of the morphogenetic tree is altered relative to the timing of links in the others.

I end this introductory section by illustrating the three possible types of change in Figure 2.1, and by making three final points. First, it is possible to compare the morphogenetic tree of one individual organism with that of another (conspecific) organism in which a mutation has occurred. Here, the nature of the change in the tree—whether phase, structural or distortional—is a *mutational* change. Alternatively, it is possible to regard conspecific individuals in natural populations as homogeneous in the nature of their morphogenetic trees, and to make interspecific comparisons (across large or small taxonomic 'gaps'). Here, the changes are *evolutionary* ones. Clearly, evolutionary changes in morphogenetic trees are based on a subset of all possible mutational ones. Second, it seems likely that many changes in morphogenetic trees (whether mutational or evolutionary) are of a 'compound' nature, that is, they may involve alterations in two or more of the three categories (phase, structural, distortional); however, it is still important, for the purposes of clear thinking, to be able to distinguish these three categories. Third, anyone familiar with the work of von Baer (1828), Haeckel (1866) and Gould (1977b) may already be able to anticipate connections between their work and my categories (von Baer—phase change; Haeckel—structural change by terminal addition; Gould's heterochrony—distortional change). These connections are discussed in the following three sections, respectively.

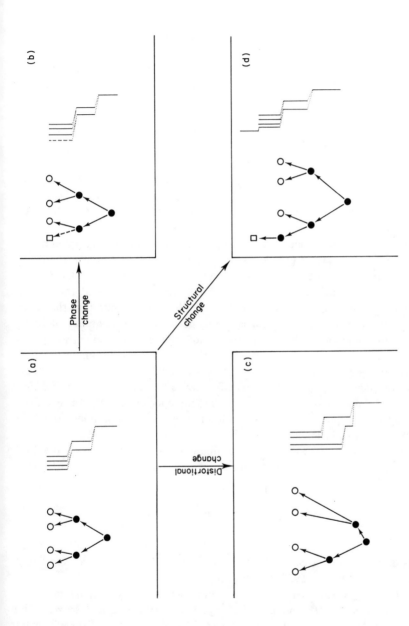

Figure 2.1 Types of evolutionary change in morphogenetic trees. (a) starting point; (b) phase change; (c) distortional change; (d) structural change. In each 'box', left hand diagram is a causal link tree, right hand diagram is a locus-based tree. • morphogenetic heterogeneity, ○ terminal heterogeneity, □ altered heterogeneity, ↑ causal link, – – – → altered instruction in 'old' causal link, ——— activity of particular locus over developmental time, – – – activity of new allele of locus. (See also Figures 1.2 and 1.5.)

2.2 PHASE CHANGE AND VON BAER'S LAW

Of the three kinds of evolutionary change which may occur in morphogenetic trees, phase change is the easiest to formalize. In this section, I will show how even a very elementary formalization produces an interesting result, namely a possible mechanism for von Baer's third law. This is the law which states that, in most interspecific comparisons, early developmental stages are more similar than later ones—a phenomenon which de Beer (1958) calls 'deviation' (in the sense of progressive deviation of form through development).

The way in which I approach phase change evolution is to consider the relative rates of substitution of new alleles at different stages of development. As in the previous chapter, I restrict the discussion to those loci (and their alleles) that have a role in controlling the developmental process (i.e. D-loci: Arthur, 1984). Genes which are active at a particular developmental stage but have no morphogenetic function, and are purely acting in a 'housekeeping' capacity (e.g. by producing a routine metabolic enzyme), are not relevant to the issue and will not be considered.

Taking any individual locus, we may define three important parameters: m, the probability of mutation (per locus per gamete per generation); a, the probability of a mutation that does occur being selectively advantageous; and k, the overall rate of substitution (i.e. fixation) of mutant alleles at that locus. I will make the assumption that a', the probability that a mutation that does occur will eventually become fixed, is the same as a, its probability of being selectively advantageous. Like many assumptions made in theoretical work, this one is patently untrue. Fisher (1930) and other authors have pointed out that many selectively advantageous alleles are lost through drift shortly after their initial occurrence in a population. This effect tends to make $a' < a$. The fixation of some neutral and nearly-neutral alleles has the opposite effect. However, the assumption makes quantification of phase change evolution simpler, and I do not think that its violation seriously affects our argument.

In a numerically static population of a diploid organism, with each individual contributing two gametes to the following generation, the rate of substitution of mutant alleles,

$$k = 2Nma \tag{1}$$

This contrasts with the simpler situation that pertains when only neutral alleles are considered, in which case mutation rates and substitution rates are equal (i.e. $k = m$; see Kimura, 1983, Chapter 3).

Corresponding to the variables k, m, and a, which apply to individual loci, are the variables K, M, and $\bar{a}$, which apply to *developmental stages*. The relationship between these two sets of variables will become clear if we let subscript i run over stages, and j run over loci *within* any particular stage. Then the overall rate of occurrence of mutations at any particular stage,

$$M_i = \sum_{j=1}^{j=n_i} m_{ij} \qquad (2)$$

where n_i is the number of loci active at the ith developmental stage. Similarly,

$$K_i = \sum_{j=1}^{j=n_i} k_{ij} \qquad (3)$$

The relationship between a and $\bar{a}$ is not quite the same. Rather, $\bar{a}_i$ represents the average probability of a mutation in a locus active at the ith developmental stage being selectively advantageous; that is,

$$\bar{a}_i = \sum_{j=1}^{j=n_i} a_{ij}/n_i \qquad (4)$$

We are now in a position to consider the substitution rates at different developmental stages. For each stage, we have

$$K_i = 2NM_i\bar{a}_i \qquad (5)$$

Taking two developmental stages, one of which ($i=t$) is earlier than the other ($i=t+1$), we have

$$K_t = 2NM_t\bar{a}_t \quad \text{and} \quad K_{t+1} = 2NM_{t+1}\bar{a}_{t+1} \qquad (6)$$

What I want to show is that, in general,

$$K_t < K_{t+1} \qquad (7)$$

Whether or not this is true depends on the relative magnitudes of M_t and $\bar{a}_t$ on the one hand, and M_{t+1} and $\bar{a}_{t+1}$ on the other. It does not depend on N, which appears on both sides of the equation.

It is perhaps worth noting, at this point, that if we were considering two stages at a considerable distance from each other in the life cycle (e.g. gastrula versus end of post-embryonic growth), then the 'population size', N, should *not* simply be inserted unchanged on both sides of the equation. This is especially true for species with a 'Type I' survivorship curve, in which mortality is greatest during juvenile stages. In such a situation, the *effective* mutation rate for the whole population, with regard to the later stage, should not be $2NM_{t+1}$, but rather somewhat less than this, because a certain fraction of mutations will be lost from the population before the loci at which they occur become active, and thus before their selective advantage, if they have one, comes into play. However, because the morphogenetic tree applies to *early* development, as noted in the previous chapter, we can be reasonably confident that this complication can be ignored, in the present argument, without significantly affecting our conclusions.

We now consider the two important parameters M and $\bar{a}$ in turn. Providing that, despite some variation in mutation rate among loci, there is no systematic temporal trend in rate according to time of activity in development, then in general,

$$M_t < M_{t+1} \tag{8}$$

simply because the number of loci is smaller at the earlier developmental stage. Even if there *is* a temporal trend in mutation rate during development (and there is currently no reason to suspect one) inequality 8 will still hold, so long as

$$\frac{\bar{m}_t}{\bar{m}_{t+1}} < \frac{n_{t+1}}{n_t} \tag{9}$$

Turning to the inter-stage comparison of $\bar{a}$, there are three different ways of making the comparison, all of which lead to the conclusion that

$$\bar{a}_t < \bar{a}_{t+1} \tag{10}$$

First of all, let us make the inter-stage comparison on the basis of 'magnitude of effect' of mutations of genes acting at the stages being compared. It is both generally accepted, and an integral part of morphogenetic tree theory as presented in the previous chapter, that mutations of genes controlling early developmental decisions will on average have larger phenotypic effects than those controlling later ones. Since magnitude of effect is negatively correlated with probability of being selectively advantageous (Fisher, 1930), this line of reasoning leads in a straightforward way to inequality 10.

The second possible way to predict the ranking of $\bar{a}$ values for different developmental stages is to construct an argument in terms of the *number of characters* affected by a mutation. If we equate the terminal heterogeneities (see Figure 1.2) of the morphogenetic tree with adult characters, it is clear that the earlier in development a gene acts, the more numerous the characters that a mutation in it will alter. The probability of improving a single character is small; and the probability of improving both of a pair of characters simultaneously is smaller still (multiplicatively smaller in the case of independent characters). The more characters affected, the smaller the probability of an overall selective advantage becomes. Thus, although there are some complications to this argument (Arthur, 1982b, 1984), and it is not as straightforward as the one based on 'magnitude of effect', it nevertheless tends to lead to the same conclusion, i.e. inequality 10.

In the two approaches described above, we have dealt with the effect of a mutation in rather non-developmental terms. Essentially, we have considered its effect on the *adult*: small versus large effect; few versus many adult characters altered. The third approach is more developmentally explicit. Here we consider the effect of a mutation in a particular causal link on those links which follow and depend upon it, i.e. subsequent causal links in the same 'branch' of the morphogenetic tree.

If, prior to any specified mutational change, an organism is reasonably well coadapted, then we may assume that late-acting loci will be fixed for alleles (or equivalence classes of alleles) which modify the effects of early-acting loci such that an integrated overall developmental result ensues. However, given a mutational change in an early-acting locus, it is reasonable to suppose that this will alter the 'preferred' allele at some, or perhaps all, *modifier* loci. (I use this old genetic term in a new and developmentally explicit sense: any locus B 'downstream' of an earlier-acting locus A in the morphogenetic tree structure is a modifier of A. Thus nearly every locus is both a 'major' locus and a 'modifier' locus depending on whether it is looked at from an upstream or downstream viewpoint.)

To have an overall selective advantage, a mutation's own direct beneficial effect on the phenotype must outweigh any developmental incompatibility it causes in its modifiers. Clearly, the probability of this happening decreases the earlier in development the locus acts and the more numerous its modifiers are. Thus, the third of our three approaches to the inter-stage comparison of $\bar{a}$ values—the one based on developmental constraint—leads to precisely the same prediction as the other two, namely inequality 10 ($\bar{a}_t < \bar{a}_{t+1}$). Note that developmental constraint, as used here, is equivalent to Riedl's (1978) 'burden' and Wimsatt's (1986) 'generative entrenchment'.

Since the two important components of the substitution rate M and $\bar{a}$ both vary in the same direction, that is, higher values for later developmental stages, it necessarily follows that inequality 7 holds, i.e. that the rate of evolutionary substitution of new alleles at early developmental stages is lower than the equivalent rate at later stages.

I now return to the connection with von Baer's third law. Since I have, throughout, been concerned solely with D-loci—those loci that jointly determine morphology—and since von Baer's law is entirely a law about morphological evolution (or rather about morphologically based natural classification when it was first proposed), it is tempting to think that we now have an obvious mechanism for the phenomenon of progressive deviation that the law describes. However, while the pattern of phase-change evolution of morphogenetic trees which I have outlined *may* constitute the mechanism behind this phenomenon, the link between the two is not as direct as might at first seem.

The problem here is that the rate of evolutionary substitution of mutations affecting development is not the same as the rate of evolutionary change of *phenotype* (at any developmental stage). Yet von Baer's law was based on purely phenotypic observations. Rates of phenotypic evolution are determined jointly by allele substitution rates and magnitude of effect of the substituted alleles. This joint production of phenotypic evolution has an inbuilt irony. Since magnitude of effect is itself a component of the substitution rate, and since, more specifically, high magnitudes of effect reduce substitution rates, developmental stages with rapid substitution rates involve less phenotypic change per substituted allele than those with lower rates. So, while phase-change evolution of morphogenetic trees may indeed produce the pattern described by von Baer's law, it will only do so providing that the differential in substitution rates between developmental stages outweighs the counteracting differential in magnitude of phenotypic effect.

There are two other possible explanations for von Baer's law. First, if there is no hierarchical causal structure of development and no developmental constraint, but merely an equal number of loci at each stage evolving with equal rapidity, progressive deviation will occur simply because at least some early-acting loci will affect both early and late phenotypic characters while late-acting loci can only affect the latter. I do not believe for a moment that this is all there is to von Baer's law, but Balinsky (1981) puts forward the opposite view.

The other mechanism producing progressive deviation is a strict neo-Darwinian one. It is possible that the versions of the morphogenetic tree based on causal links and loci are correct, but that the mutational version is incorrect because all loci at all developmental stages can experience very small-effect mutation. Natural selection may then be 'blind' to the systematic trend in *average* magnitude of effect in development because micromutations are available at all stages and all evolutionary changes proceed through them. If this is so, then early stages will still experience lower substitution rates—if only because they have fewer loci—and this differential will *not* be counteracted by an inverse differential in magnitude of effect.

It is impossible to choose between these three mechanisms for von Baer's law. Of course, the one favoured by Balinsky (1981) must provide at least part of the explanation, and so it is not a straight choice between three mutually exclusive possibilities. But this statement is as far as we can go at present.

I do not wish to end the present section on this rather negative note, and there are two more positive points that should be made here. First, the fact that we cannot yet be certain which is the correct explanation of von Baer's law should not stop us probing this important question. There is already too long a history of uncritical acceptance of this law unaccompanied by enquiry into *why* the progressive deviation it describes is generally found (though see Bonner, 1974, pp.56–57; and Gould 1977b, p.231) for a more enlightened approach). For example, de Beer (1958) clearly divorces the *causes* of evolution (which he gives as mutation and selection) from its morphological *consequences* (such as progressive deviation), and seems to see no need to look for mechanisms underlying those consequences in terms of interactions between selection and developmental constraint (or in any other terms for that matter). The transformed cladists, in simply accepting von Baer's law and using it as a means of polarizing homologies (see, for example, Patterson 1982), are unfortunately continuing this tradition. Despite the existence of exceptions to it, von Baer's law is one of the most striking generalizations about morphological evolution, so failure to enquire what mechanism underlies it seems incomprehensible.

Second, looking at progressive deviation as a consequence of phase-change evolution of morphogenetic trees actually reduces the number of apparent exceptions to it. A great class of exceptions—larval adaptation or 'caenogenesis'—largely disappears when we appreciate that von Baer's law should apply *separately* to the different morphogenetic trees of a complex life cycle such as that of the holometabolous insects.

2.3 STRUCTURAL CHANGE AND MORPHOLOGICAL COMPLEXITY

In retrospect, one of the main flaws in my earlier treatment of morphogenetic tree evolution (Arthur, 1984) was that I concentrated on phase changes (Chapter 12) to the virtual exclusion of structural changes (and distortional changes). This omission has been pointed out by Garcia-Bellido (1985), who has correctly criticized me for not considering 'the possibility that "major" functions in derived forms could have been "minor" ones early in evolution.' I now rectify this omission.

In fact, Garcia-Bellido's comment is very similar to one made by Mather (1941), who suggested that 'major, switch or oligogenes were at one time polygenes'. When suggestions of this kind are considered in terms of the evolution of morphogenetic trees, it is clear that the issue is the 'growth', in an evolutionary sense, of these trees. If new causal links are added to the end of a morphogenetic tree, each previously existing link becomes more developmentally constrained, and more 'major', than it used to be.

We are in the process of arriving, here, at one of the most important, yet least understood, of all evolutionary questions, namely: how are complex organisms produced from simple ones? There is no need to pretend that this question is somehow invalid because we cannot adequately measure organismic complexity. There is little doubt that eukaryotes are more complex than prokaryotes, metazoans more complex than protozoans, and *some* later-evolved metazoan groups (e.g. insects, vertebrates, molluscs) more complex than *any* of the earliest (Precambrian) metazoa. The production of more complex organisms over *long* periods of evolutionary time can hardly be disputed. However, although this is indeed so, there is no *universal* trend towards increased complexity—just as there is no universal evolutionary trend towards increased order, organization, fitness or anything else. Decreases in complexity can and do occur, and indeed we shall see later (next chapter) that states of novel complexity may often have to be reached by a prior descent into relative simplicity; we thus need to understand 'downward' changes as well. But even so, the evolution of the more complex from the less complex constitutes one of the foremost challenges to the evolutionary biologist, representing, as it does, the origin of novelty.

A cautionary comment is necessary at this point: we have to be careful about equating complex adult morphology with complex morphogenetic trees. I suspect that these will usually go together—more causal links producing a more complex adult morphology—and I will generally make this assumption in the following discussion. However, there may be some situations in which this correspondence breaks down.

The idea that morphogenetic trees usually grow by terminal addition of new causal links is a plausible one, if only because the alternative, that is, incorporation of new links at early or intermediate stages involving major developmental decisions, seems very unlikely. However, even the less drastic form of growth (i.e. by terminal addition) requires new genes to control the new causal links. (Sometimes old genes may exert such control by acquiring a new function, but this cannot always be the case, since genome size has generally increased in evolution.) Where do the new genes come from? Among the most probable sources is gene duplication and divergence, as discussed at some length by Ohno (1970). As Ohno suggests, a redundant second copy of a duplicated gene may escape the usual rigours of purifying selection; and in a few such cases, the accumulation of normally 'forbidden' mutations may lead to a new function for the gene concerned.

Anyone familiar with the outcome of the conflict between von Baer's law and Haeckel's (1866) 'recapitulation', as presented by de Beer (1958) and Gould (1977b), may have become slightly uneasy because growth of morphogenetic trees by terminal addition provides a mechanism for recapitulation, which is now generally rejected. However, Haeckel's recapitulation captures the essence of an evolutionary phenomenon that von Baer's law does not capture, namely, the production of complex organisms from simple ones. The usual example of von Baer's law involves a comparison of the embryos of reptiles, birds and mammals, i.e. organisms at about the *same* point on the scale of complexity. In this case, progressive deviation, not recapitulation, is what we see. But if we make a broader-scale comparison (e.g. of solitary flagellate, colonial flagellate, flatworm, lamprey and mammal), the situation is hardly describable *solely* in terms of progressive deviation. Of course, the embryo of each group does not precisely resemble the adults of those 'below' it, as Haeckel (1866) seems to have suggested, and I address, below, the question of why they do not. Nevertheless, von Baer's law, while generally true, does not deal with the evolutionary increase in complexity. In this sense, the general von Baerian and Haeckelian ways of looking at evolution are complementary, rather than contradictory, and should both be retained, albeit we cannot accept recapitulation in the form that Haeckel proposed.

Why is it that strict Haeckelian recapitulation does not occur, if growth in morphogenetic trees normally takes the form of terminal addition? Basically, because the other kinds of morphogenetic tree evolution are going on alongside such growth. In particular, phase change is likely to be much more common, and therefore operating on a shorter evolutionary timescale, than structural change (since it involves mutation of existing loci rather than production of new ones). So by the time new links are added to the morphogenetic tree in a particular phyletic line, phase changes in that line and any other chosen for comparison will have caused a considerable degree of von Baerian deviation.

The opposite to 'growth' of morphogenetic trees by addition of new causal links (terminal or otherwise) is simplification of the tree structure by loss of links. It is difficult to assess the relative frequency of these two types of structural change for the following reason. The rate of simplification is proportional to the product $M_E \bar{a}_E$, where M_E is the overall rate of occurrence of mutations that *eliminate* causal links, and $\bar{a}_E$ their mean probability of being selectively advantageous. (These parameters are 'overall' ones in the sense that they are restricted neither to any locus nor to any developmental stage.) Similarly, rate of growth is proportional to $M_A \bar{a}_A$, where subscript A indicates addition of links. The literature on morphological mutants of *Drosophila*, as summarized by Lindsley and Grell (1968) suggests that loss of links (e.g. eyeless, wingless, etc.) is much more common than addition (a possible example is multiple wing hairs); that is,

$$M_E \gg M_A \qquad (11)$$

and it seems intuitively reasonable that this should be so. On the other hand, the loss of causal links in a well-established, coadapted morphogenetic tree system would be expected to be selectively disadvantageous in the vast majority of cases, while addition of links—especially terminal ones—might not be detrimental nearly so often.

Thus we might expect that

$$\bar{a}_E \ll \bar{a}_A \qquad (12)$$

Given these two counteracting inequalities, the relative frequency of growth and simplification of morphogenetic trees can hardly be predicted; nor does the available evidence on morphological evolution allow an easy assessment of which has been more common. Of course, the most complex living organism has tended to ascend the scale of complexity during the course of evolution, while the least complex has stayed more-or-less constant; but this in itself does not tell us much about the relative frequencies of the two kinds of evolutionary event. Even if, as I have suggested earlier (Arthur 1984, Chapter 13), the 'species of median complexity' moves up the scale during evolutionary time, this still does not require an excess of 'growth events' over simplifications.

I should stress that structural changes more complicated than simple addition and elimination of links are possible. For example, if a sequence of links A → B → C is replaced by A → C, we have a structural change which can be thought of as a composite of two deletions and one addition. Pritchard (1986, Chapter 15) has argued for the importance of changes of just this kind in evolution. While no doubt some such changes have occurred, I cannot currently see any useful way of generalizing about them, and so I will not devote any further space to this issue.

Finally, if elimination of late causal links is more common than elimination of early ones, as seems likely, then both primary types of structural change of morphogenetic trees—growth and simplification—parallel phase change in being subject to a degree of constraint that varies systematically in developmental time.

2.4 DISTORTIONAL CHANGE AND HETEROCHRONY

Both phase change and the 'growth' variety of structural change of morphogenetic trees involve the transmission of new developmental messages. In phase change, we have new messages in old causal links; in 'growth', we have new messages in new causal links. Simplification of trees (the alternative kind of structural change to growth) involves *loss* of messages. However, it is also possible for morphogenetic trees to evolve without either gain or loss of messages, simply by alteration of the timing of causal links in one or more branches of the tree relative to others. This process, which I call distortional change (see Figure 2.1), is, with a qualification to be discussed below, equivalent to heterochrony seen in a morphogenetic tree context. It may be 'inter-somatic' heterochrony, where one somatic part (e.g. an organ system) is accelerated or retarded in its developmental production relative to other such parts; alternatively it may be 'germ-soma' heterochrony, where the germ line is accelerated or retarded relative to the development of somatic organs generally.

Heterochrony has been dealt with very thoroughly by Gould (1977b), and I will only make two brief points here. First, continuing the theme of developmental constraint, we should expect distortional changes in late developmental stages to be commoner than those in early ones. Second, germ-soma distortional change in which the germ-line is accelerated relative to the soma may often be accompanied by structural change because terminal somatic causal links may be entirely lost. Indeed, neoteny and progenesis are normally just such a combination of distortional and structural change, rather than being purely distortional. As noted by many earlier authors, including de Beer (1958: 'clandestine evolution') and Gould (1977b), these processes may represent an important route to novel morphologies, a point to which I will return in the following chapter.

2.5 SUMMARY

Building on the concept of the morphogenetic tree formulated in the previous chapter, three types of evolutionary change can be distinguished: phase change, structural change and distortional change. Phase changes involve altered developmental instructions being transmitted in an *unaltered* tree structure. Structural changes involve addition and/or deletion of causal links. Distortional changes involve altered timing of causal links in one or more branches of the morphogenetic tree relative to others.

With regard to these three kinds of morphogenetic tree evolution, I make the following proposals:

1. All three types of change are subject to increased constraint in earlier developmental stages.
2. This temporal trend in developmental constraint, particularly when applied to phase change, is partly responsible for the phenomenon of progressive deviation described by von Baer's third law.
3. The production of more complex organisms from simpler ones over *long* periods of evolutionary time is based largely on structural change through 'terminal addition' of causal links.
4. However, this will not produce strict Haeckelian recapitulation, largely because of phase changes acting on a shorter timescale.
5. 'Simplification' of morphogenetic trees (the subcategory of structural change in which causal links are lost) combined with distortional change to accelerate the production of the germ-line relative to the soma is one important route to novel morphologies, because it provides a means of reducing developmental constraint.

The last of these proposals—along with other routes to novel morphologies—will be considered in the following chapter.

Chapter 3

Mechanisms for Major Evolutionary Transitions

3.1 Introduction 34
3.2 Primary divergence 38
3.3 Escape to simplicity. 39
3.4 Morphological windows 41
3.5 Escape from competition. 44
3.6 Summary and conclusion 48

3.1 INTRODUCTION

In the previous chapter I put forward what might be described as the general trends of morphogenetic tree evolution: more rapid evolution of terminal causal links than earlier ones, phase change more rapid than structural change, and so on. Such trends provide a possible explanation for certain widely observed patterns in morphological evolution, including von Baer's progressive deviation. However, the previous chapter did not deal explicitly with the origin of major morphological types (body plans or Baupläne) as represented by the higher taxonomic groups, from about the family/order level to the phylum and above. This omission was deliberate. The issues of how major body plans arise, whether they require any special explanation over and above accumulated intraspecific divergences and speciations, and indeed whether there is any such thing as a body plan, are highly contentious. They require separate and direct consideration, which is the aim of the present chapter.

I start by noting that I use the term 'body plan' in a relative way, and do not recognize any particular level of decision within the morphogenetic tree structure whose evolution equates with a switch in body plan. The decisions at the very base of the tree are clearly involved in establishing the basic body layout; terminal decisions at the top of the tree have relatively little morphological effect; in between, there is a gradual diminution, going 'upwards' through the tree, in the extent to which evolutionary change in the nature of the developmental decisions concerned can be considered as body plan changes. This 'continuous', rather than discrete, view of development, which is one of the strengths of the morphogenetic tree concept, turns up in various other guises. We have already seen it in the new interpretations of 'major gene' and 'modifier' in the previous chapter; and it will show up yet again below in the context of what constitutes a macromutation.

This relative, rather than absolute, usage of 'body plan' might give the impression that what follows will be along strict neo-Darwinian lines, inasmuch as neo-Darwinism and 'discrete body plan' views of evolution are generally regarded as being opposed. However, as will be apparent from the previous chapters, morphogenetic tree structure, as I have presented it, does not permit the total organismic plasticity that the strict neo-Darwinian view requires. Rather, below a certain level in the tree structure, many genes with a developmental effect are viewed as being highly constrained, incapable of effective micromutation, highly conserved evolutionarily, and capable of evolving only through very rare advantageous macromutations.

At this stage, we need to have working definitions of the terms 'micro-' and 'macro-' mutation so that we can start to use them in a more precise way. Related to this, we also need a clearer idea of what is meant by neo-Darwinism. I will take these two issues in turn since obviously we cannot 'define' neo-Darwinism if we do not know what constitutes a micromutation.

From various experimental studies on the effects of mutations on the phenotype, we know that mutations range from those whose effects are individually undetectable (the classical province of quantitative genetics : see Falconer, 1981) to those which involve radical morphological change (best known in *Drosophila* : see Mahowald and Hardy (1985) for a recent review). Lying between mutations which, acting alone, affect whole body regions, and those which, acting jointly, add a single abdominal bristle, there is a more-or-less continuous spectrum of mutations with almost every conceivable magnitude of morphological effect. In morphogenetic tree theory (see Chapter 1), this variation in the magnitude of effect of mutations is seen as having both within- and among-locus components; and the latter is seen as having within- and among-developmental stage subcomponents.

Clearly, both observation and morphogenetic tree theory argue against a clean separation of two alternative categories of mutation—micro and macro. Nevertheless, it is useful to be able to split the continuum of 'magnitude of effect' at some (necessarily arbitrary) point, so that we can talk loosely about micro- and macromutations, and thus make a connection with the numerous earlier authors in whose work these two terms are central. I will thus adopt the following distinction. Micromutations are those whose effects on morphology are individually indiscernible. They are mutations which would not lend themselves to Mendelian breeding experiments, because we could not recognize morphological categories among which to observe offspring ratios. Their products are typified by the alleles underlying continuous morphological variation. All else is macromutation.

I make the assumption (see Chapter 1) that, in a morphogenetic tree system, genes contributing to the latest developmental stages are capable of micromutation, thus 'defined', while the earliest stages involve some genes which are not. The level below which one begins to encounter genes in which micromutation is not possible, going backwards through developmental time, is not yet identifiable, but presumably varies among different kinds of organism.

In case the above discussion serves to create the impression that 'polygenes' and genes making the terminal decisions of the morphogenetic tree are one and the same, let me add the following comment. The morphogenetic tree is, as I have stressed earlier, an abstraction of *early* development, during which 'characters' are being formed. This early formative period is gradually replaced by a period of post-embryonic growth, in which already-formed characters simply grow, albeit usually allometrically in relation to each other. Polygenes probably include both those genes controlling the terminal causal links of the morphogenetic tree *and* those controlling post-embryonic growth (e.g. by being involved in the production of agents such as growth hormone).

Having arrived at a working definition of 'micromutation' and 'macromutation', it should now be possible to identify the fundamental nature of neo-Darwinism. We immediately meet a problem, though, because, as noted by several authors including Waddington (1975), there are two 'neo-Darwinisms', which I will call broad and strict. Broad neo-Darwinism, as exemplified by Maynard Smith (1969), is basically the non-Lamarckian, Darwin–Mendel–Weismann view of the living world, with, I would add, though Maynard Smith (1969) does not, an emphasis on the *predominance* of micromutational change in evolution. I use the term 'modern synthesis' synonymously with broad neo-Darwinism. Strict neo-Darwinism, on the other hand, allows *no* macromutational input into evolution. Darwin himself, although a pluralist on so many other evolutionary issues, often advocated the 'strict' approach, with his repeated statement (1859) that 'Natura non facit saltum', and was criticized by his otherwise supporter T.H. Huxley (See Darwin, 1887) for doing so. Mayr (1963) illustrates a similar 'strictness' when he states that '*all* evolution is due to the accumulation of small genetic changes, guided by natural selection' (my italics). (It is in fact clear from the context that Mayr means micromutations, i.e. genetically-based small *phenotypic* changes, which is of course the real issue, and not small *genetic* ones; there is not a direct correspondence between these two.)

The relationship between the views put forward herein and neo-Darwinism (which will be considered in more detail in Chapter 4), entirely depends on which of the two versions of neo-Darwinism is involved. I see morphogenetic tree theory as an attempt both to visualize in more detail, and to explain, the pattern of predominant micromutational change and occasional macromutational change that is implicit in broad neo-Darwinism. On the other hand, the theory of morphogenetic tree evolution, as presented in this book, is *completely incompatible* with strict neo-Darwinism. These two views are mutually exclusive: either (or neither) may be a correct view of the living world, but both of them cannot be.

Of course, neo-Darwinism, in either form, is not characterized solely by its emphasis on micromutation. Other key features include Mendelian rather than Lamarckian inheritance; natural selection rather than other evolutionary mechanisms; and selection at the level of the individual rather than at other levels. While there are those who would have us question the first of these features (Gorczynski and Steele, 1980), and while the other two are under more credible attack, albeit in restricted provinces (Kimura, 1983; Stanley, 1975, 1979, respectively), they are not at issue herein, and consequently I will not devote any further discussion to aspects of neo-Darwinism other than its emphasis (exclusive or otherwise) on micromutation.

Another continuum that is often broken up into categories for ease of discussion is that of micro-/macro-/mega-evolution. As before (Arthur, 1984), I follow Simpson (1944) in using microevolution for intraspecific changes, macroevolution for the origin of species and genera, and megaevolution for the origin of higher taxa (from about families/orders upwards). One difficulty with these terms is that they may be used in relation to magnitude of evolutionary change or to timescale (roughly, hundreds vs millions vs hundreds of millions of years), or to both. I will try to make it clear, when I use these terms in any particular context, which sense is implied. For the moment, I will simply note that morphogenetic tree evolution is a pattern which is observable in its entirety only on the megaevolutionary timescale. This is in contrast with, for example, punctuated equilibrium (Eldredge and Gould, 1972; Gould and Eldredge, 1977) which is a pattern in macroevolutionary time, and the evolution of industrial melanism (see Bishop and Cook 1980), which is a microevolutionary phenomenon.

If the general pattern of morphogenetic tree evolution described in the previous chapter is correct, can we not: (a) simply equate the predominant micromutational changes with common microevolutionary transitions and the rare macromutational shifts with major evolutionary transitions; and (b) just accept that in some rare situations there are certain special 'one-off' reasons why mutations with large effects are indeed advantageous? In fact, the answer to both parts of the question is 'no'. First, as I will argue in the following sections, macromutations can constitute no part, a small part, or a large part of the cause of a major evolutionary transition. Second, while occasional advantageous macromutations may indeed sometimes occur for unique reasons, there may nevertheless also be certain circumstances—both external, ecological ones and internal, developmental ones—which make advantageous macromutation more likely. If such sets of circumstances can be identified, or at least postulated in a plausible way, then we are a little further on than we were when simply making the bald assertion that some mutations with large phenotypic effects will convey a selective advantage.

The following four sections describe four modes of production of major evolutionary transitions which are compatible with morphogenetic tree theory. Macromutations are not involved in the first, are a probable but non-essential part of the second, and necessarily constitute the main stimulus for change in the third and fourth. In the first three, the concept of developmental constraint is central and the approach taken is largely 'internal'; in the fourth, the emphasis is on overcoming 'ecological constraint' and the focus is on external, niche-related factors.

Finally, before leaving this introductory section and proceeding to the discussion of these four mechanisms, it might be useful to clarify what I mean by the term 'major evolutionary transition' used in the title of this chapter. I use this phrase in an all-inclusive way for the evolutionary production of the difference between any two major body plans, as represented by higher taxa such as classes or phyla. Specifically, I include both cases in which the difference is achieved gradually and those in which it is achieved suddenly. Also, I include all degrees of asymmetry of divergence, from cases in which one of the groups being compared has diverged little from the common ancestor, the other a lot, to those in which the degree of divergence is approximately equal.

3.2 PRIMARY DIVERGENCE

Let us consider the evolution of two present-day higher taxa of complex morphology from a common ancestral form of much simpler morphology in the distant evolutionary past. For example, if we accept, as a basis for discussion, Barnes' (1980) 'possible phylogeny of the Animal Kingdom', then both vertebrates and insects are ultimately derived from a flatworm of some sort; and indeed, an ancient flatworm was their *most recent* common ancestor.

Because we are dealing here with very long periods of evolutionary time, over which a considerable increase in morphological complexity has taken place, there must, over this period, have been substantial *growth* of morphogenetic

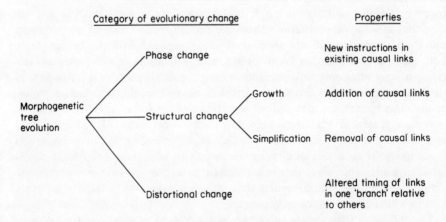

Figure 3.1 Types of evolutionary change in morphogenetic trees

trees, as well as phase and distortional changes (see Figure 3.1 for a summary of the meanings of these terms). Because of the problem of developmental constraint of early embryological stages and the genes controlling them, this growth will largely take the form of terminal addition of causal links, as discussed in the previous chapter.

As terminal links are added, all earlier links become more constrained, and more 'major' in a passive way, simply due to their coming to control more and more later decisions. I have earlier pictured this as a gradual shifting of the whole tree structure—expressed in its genetic form—to the right, so that the latest links always occupy the same position on the 'magnitude of effect' axis (Arthur, 1987a, Chapter 12).

If, at any particular stage in the evolutionary divergence of the lines leading to insects and vertebrates, all mutational changes being incorporated are late-acting micromutations, then the process by which the major taxa arise is a strict neo-Darwinian one. The 'gap' in body plan arises gradually, by an accumulation of many small changes—a mixture of phase, structural and distortional ones—over a very long period of evolutionary time.

While doubtless much of the morphological gulf between any pair of existing higher taxa is due to this 'primary divergence', there is a problem which faces anyone proposing that this is all that there is to the production of groups with radically different body plans, such as, in our example, insects and vertebrates. The problem is that on its own this mechanism will produce recapitulation of early stages, i.e. embryos resembling adult flatworms in all derived complex forms. Although there are some striking examples of evolutionary conservation of very early developmental processes, such as the similarity of early embryogenesis in molluscs and annelids (summarized in Slack, 1983, Chapter 5), there has also, in the long term, been considerable *change* in early development, even between higher taxa recognized as being 'close' evolutionarily. The contrast between annelid spiral cleavage and insect syncytial cleavage is a case in point.

What this means is that unless many metazoan phyla share no multicellular ancestor, or unless even the very earliest developmental stages can somehow evolve by micromutation (which in morphogenetic tree theory is taken to be impossible), evolutionary production of today's phyla cannot be achieved by primary divergence alone. The following three sections consider mechanisms which, superimposed on primary divergence, can produce a better explanation of the observed pattern of morphological relationship among extant groups.

3.3 ESCAPE TO SIMPLICITY

In the process of primary divergence leading to two or more higher taxa which are more complex morphologically than their common ancestor, positive structural change of morphogenetic trees (*growth*) outweighs its negative equivalent (*simplification*). (See Figure 3.1 for a reminder of the meanings of these terms, and Chapter 2 for their original definitions.) Thus, as the process continues, there is a general accumulation of developmental constraint, and therefore a general decrease in the potential for radical morphological change. One possible escape route from this build-up of constraint is a lineage in which a sudden and considerable simplification takes place, e.g. a removal of the terminal level of causal links, together with several immediately underlying levels. Effectively, the top of the morphogenetic tree is 'chopped off'. As major changes go, this kind of change should be *relatively* unconstrained, because the juvenile form which now becomes the adult has already had a long history of selection. It should thus already be a well-coadapted entity in its own right, and, if it is an *active* larval or juvenile form, then it will also be adapted to a particular ecological role. This is in marked contrast to the problems that arise in a major mutational restructuring of an already-existing adult to produce, at once, a radically new adult.

The main problem that *is* met with, of course, is that to become adult in the first place, our simplified juvenile form must become sexually mature. Thus, the branch of causal links leading to the germ-line must be accelerated in relation to the rest of the morphogenetic tree, which produces somatic structures. This combination of tree simplification and distortional change involving relative acceleration of the germ-line corresponds, as I pointed out in the previous chapter, to the processes of neoteny and progenesis.

Gould (1977b) has discussed the evolutionary significance of these processes at great length, so I will restrict the discussion here to a re-emphasis of three of the points Gould makes which seem to me to be particularly important. First, the long-term escape from developmental constraint and consequent potential for new morphological direction that these processes can produce is a *result*, which tells us little about the selective *cause* of the initial jump 'backwards' into relative simplicity. Probably, at least in the case of progenesis, the cause is a local environment that favours high reproductive potential. That is, for an explanation of a process with long-term morphological consequences we look, as usual, to short-term ecological advantages: in this case, the advantages of certain life histories (or, if you like to think of organisms planning their own evolution, 'life-history strategies').

Second, progenesis and neoteny need not necessarily be achieved suddenly, in a macromutational manner. Indeed, Gould states (1977b, p284): 'Neoteny has also been favoured because it provides one of the few mechanisms for rapid and profound evolutionary change in a Darwinian fashion without the specter of macromutation.' However, I re-emphasize my own earlier point that macromutational production of life-history changes is less fraught with problems than other sorts of macromutation.

Finally, as Gould stresses, progenesis (precocious maturation) may have different evolutionary consequences from neoteny (retardation of somatic morphological development), and it is the former that is likely to be more important in the origin of major taxonomic groups. This is partly because it frees more of the genome from its previous developmental function, and consequently releases a large potential store of genes which may accumulate previously 'forbidden' mutations, and acquire new roles, in the same way that characterizes Ohno's (1970) duplicated genes discussed in the previous chapter.

3.4 MORPHOLOGICAL WINDOWS

It seems unlikely that complex organisms cannot make major morphological changes without first returning to a state of relative simplicity, and that all evolutionary change must either take the form of primary divergence (from a primitively simple state) or 'secondary divergence' (from a derived, progenetically simple state). This section and the next are concerned with two possible mechanisms for the evolutionary production of one complex organism from another, very different, complex organism. The 'morphological windows' discussed in the present section involve mutations which, though macromutations, are selectively advantageous in a standard Darwinian way; in the following section I discuss the possibility that evolution may incorporate mutations which are in a sense highly detrimental.

What I mean by a 'morphological window' is as follows. There may be an ensemble of morphological states each reachable from all other states in the ensemble through quantitative modification, but separated by a 'gap' in morphospace from a second ensemble of the same sort. (Sinistral and dextral gastropods are examples of such ensembles.) If transitions between one particular state in the first ensemble and one in the second are possible by selectively advantageous macromutation, then those two states between which macromutational change occurs each constitute morphological windows to/from their ensemble. Clearly, ensembles may have more than one window. Also, there may be macromutational windows between non-disjunct areas of morphospace, where a lengthy micromutational transition between the states at opposite ends of the window could eventually achieve the same result (e.g. changes in arthropod segment number).

Although our knowledge of morphology is too rudimentary to allow accurate prediction of which morphological states will constitute windows, there can be little doubt that a central issue here is the degree of independence of different morphological characters; and a discussion of this is now necessary, starting with the question of what 'independence' means in this context. Of course, it is clear that all the phenotypic effects of an individual mutation are ultimately connected, even if the causal pathways leading to them diverge at the immediately post-genic stage. Nevertheless, we may recognize a sort of 'effective independence' in the following way. In any one body region affected by a mutation, the alterations to different components in that region are highly integrated and non-independent. An example frequently given is the mutation causing polydactyly in man, where the extra digit has its extra nerve and blood supply as well as its own skeletal, muscular and skin components, all forming a fairly normal pattern. Clearly, while this mutation is disadvantageous, it would be much more so if the different components were rearranged in a non-integrated, effectively independent way. In contrast to this, a mutation's effects on different body regions or on characters transcending individual regions may be 'effectively independent' in that the compounding of the directions of effect on the different characters may be such as to increase *or* decrease the overall fitness of the mutant form. Mutant wingless *Drosophila* are markedly less fit than the wild type (except, perhaps, on a windswept oceanic island), because of their lack of flight. They are also, other things being equal, lighter. This may sometimes be selectively advantageous, and other times disadvantageous, but which it turns out to be in any particular case is effectively independent of the nature of the primary effect on the wings.

I would like to examine a comment made by Erwin and Valentine (1984) which either clarifies or confuses the issue of independence, depending on how it is interpreted. Erwin and Valentine suggest that early metazoans might have been more morphologically 'compartmentalized'. If so, they argue, 'major mutations would have affected individual components independently and would have a better chance to be adaptive'. If they mean (as I think they do) that a mutation affecting one character *only* has a higher probability of being selectively advantageous than one which affects several characters *independently*, then it is difficult to disagree with them. (Indeed, this argument is similar to one I used earlier: see Chapter 2.) If, on the other hand, they mean that a mutation affecting n characters *independently* has a greater probability of being selectively advantageous than one which affects n characters *non-independently*, then I think that not only are they wrong, but that precisely the opposite is true.

What I wish to propose is that the existence of morphological windows will be favoured at both extremes on the spectrum of independence: characters or character-complexes which can be radically altered by mutation without ramifications for other characters (maximal independence); and organismic states where *all* characters can be altered in a co-ordinated, non-independent way by a single mutation (maximal non-independence).

An obvious example of maximal independence is pigmentation. Although this is not part of morphology *sensu stricto* (see Arthur, 1984, Chapter 1), it is sometimes considered to be a morphological character in a broad sense of the term, and it is instructive at any rate to consider the *contrast* between the evolution of pigmentation and the evolution of 'strictly' morphological characters such as size and shape. In particular, it is clear that the frequency of macromutation is orders of magnitude higher in the evolution of pigmentation than in the evolution of organismic structure; and the explanation of this must, I think, be sought in terms of the relative independence of pigmentation from other characters. It is possible to radically alter the external pigmentation over large areas of a butterfly, moth or snail by a single macromutation, and yet create virtually no coadaptational problems. Such macromutations have frequently contributed to evolution in the groups mentioned, as is widely known. In contrast, severe coadaptational problems would arise from a macromutation whch altered the size or shape of a particular organ, which presumably is why such macromutations contribute much more rarely to evolution.

Macromutations of pigmentation (for example, those in *Cepaea* reviewed by

Jones *et al.*, 1977) can be considered as morphological windows—or, perhaps better, pigmentation windows—in the following way. Consider the base colour of a *Cepaea* shell (as opposed to its banding). Alternative alleles at a major colour locus give different colour categories : essentially yellow, pink and brown, though the situation is somewhat more complicated than that. Each of the categories produced by the major colour locus is in fact a large series of intergrading shades, although *Cepaea* workers have not emphasized this fact. Thus each major colour category is an ensemble of shades, and mutation at the colour locus makes the transition from one ensemble to another. However, no one particular state (i.e. of shade) is necessary before the transition can be made to a different ensemble (i.e. of colour). There is thus, in a sense, a *generalized* window.

Morphological windows associated with mutations affecting *all* characters simultaneously may be exemplified by the switch between sinistral and dextral gastropods. This switch can occur in a vast array of gastropod species (Pelseneer, 1920), yet within a species the rarer type is present as more than a very occasional variant only in *Lymnaea* (Boycott et al., 1930), *Laciniaria* (Degner, 1952) and *Partula* (Clarke and Murray, 1969). Also, most gastropod families are uniformly or near-uniformly dextral, while others, such as the Clausiliidae, are essentially sinistral. This suggests two things : first, evolutionary change by macromutation is possible; second, it occurs with extreme rarity.

If we consider the vast range of shapes and sizes of dextral shells as one morphological ensemble, and the equivalent range of sinistral shells as an alternative one, then we obviously have a situation where quantitative modification is possible within ensembles but not between them. The jump from one to the other *must* be made suddenly; and I would postulate that it normally arises macromutationally or ecophenotypically, not via quantitative polygenic modification.

What is not clear is whether the dextral and sinistral ensembles are connected by a diffuse, generalized morphological window, or by a few specific windows. The question is basically this : are the few cases in which an evolutionary switch in chirality has been made determined by a particular (for example) dextral morphology in which the switch to sinistrality is particularly easy; or were they simply determined by unique ecological circumstances which could just as easily have happened to snails of a different dextral morphology? The first possibility implies specific windows, the second a generalized one. I do not think we can currently choose between these explanations; though if the ramifications of a switch to sinistrality discussed by Gould *et al.* (1985) vary in magnitude among different dextral morphologies, as we might expect them to, then this would suggest *specific* morphological windows.

The existence of morphological windows associated with co-ordinated macromutational change of *all* characters, as opposed to restriction of such change to a single, independent character, has an implication for the shape of the distribution of 'probability of a mutation being selectively advantageous' against 'magnitude of phenotypic effect'. It is implicit, both in broad and strict neo-Darwinism and in the general pattern of morphogenetic tree evolution discussed

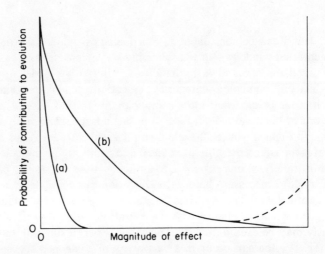

Figure 3.2 Sketch of the dependence of 'probability of contributing to evolution' on 'magnitude of effect' under (a) strict neo-Darwinism; (b) broad neo-Darwinism/morphogenetic tree theory without (solid tail) or with (dashed tail) modification suggested by consideration of the non-independence of mutational effects on different characters. (y axis is non-linear.)

in the previous chapter, that this distribution is 'J-shaped'—actually, a laterally inverted J with a flattened tail—see Figure 3.2. (The difference between broad and strict neo-Darwinism is in the length of the tail.) However, the above considerations about the degree of independence of characters suggest a U-shaped component (see Figure 3.2). The alternative possible mechanism for major evolutionary transitions discussed in the following section also suggests such a component, though as we will see, for a very different reason.

3.5 ESCAPE FROM COMPETITION

As well as being constrained developmentally in their evolution, most species are 'ecologically constrained' in as much as their niches are surrounded in ecospace by the niches of other coexisting species. Complementary to this picture of evolution being blocked by actual or potential interspecific competition is a picture of opportunity for new evolutionary direction when an adjacent area of ecospace is unoccupied. This connection between lack of competition and new evolutionary direction has been made, in a general way, by a great many authors (e.g. Simpson, 1944, 1953; Stanley, 1979; Arthur, 1987b). When the 'new direction' is embarked on by utilization of micromutational variation, the idea meets with little resistance among evolutionary biologists. While such a process is no doubt quite common in evolution, I have said enough about it elsewhere (Arthur, 1987b) and will not devote any space to it here. Rather, this section is concerned with describing a possible connection between empty ecospace and *macro*mutational change, which must be very rare, but, I would argue, of considerable importance when it does occur. For reasons to be discussed below, I have called this mechanism for establishment of occasional macromutational variants *n*-selection (Arthur, 1984).

The difference between *n*-selection and standard Darwinian selection is based on the criterion for success or failure of any particular variant. In Darwinian selection, a variant 'succeeds' if, for that variant, fitness, as measured by the cross-product ratio w, is greater than 1 (see, e.g., Cook, 1971, Chapter 1). In *n*-selection, the criterion for success is, for our hypothetical variant, $R_0 > 1$. R_0 is the 'net reproductive rate' of life-table theory, as described by Krebs (1985), applied here to the mutant subpopulation alone.

Because *n*-selection is so fundamentally different a process to what we normally consider to be selection, it is helpful to illustrate the process by use of a gedanken experiment. Reproduced below is such an experiment, as described in Arthur (1984, pp.183–184). The mutations mentioned in the quoted passage are all macromutations, as defined in the first section of this chapter.

> Let us suppose that a series of population cages of *Drosophila melanogaster* is set up, and into each is placed a propagule of, say, 100 fertilized eggs of a particular genotype. Each cage is inoculated with a different genotype, but each genotype is a homozygote for a mutant allele that affects morphogenesis is some fairly substantial way.... We then ask the question: Which of the genotypes will, in these pure cultures, establish persistent and presumably stable populations? It is easy to find examples of mutations for which the outcome of such a culture is known. The vestigial-winged homozygote *vg/vg* would establish a stable population; the larval lethals studied by Nüsslein-Volhard and Wieschaus (1980) clearly would not. Although a few genotypes would no doubt give indeterminate results in that some replicate propagules would establish populations and others would not, due to uncontrolled environmental variables or variation in the genetic background, the vast majority of morphogenetic mutants could be classified as definite successes or failures. The criterion of success in such an experiment is that, *at low density*, $R_0 > 1$. (It is assumed that the cages have sufficient resources so that $n=100$ of any life-stage is substantially below carrying capacity.) In the case of a population of asexual organisms with no age structure, this simply means that when resources are plentiful the average number of offspring per individual exceeds 1.0.
>
> It is clear that in our overall gedanken experiment a form of selection has occurred: some variants are fitter than others (as measured by R_0), and this variation in fitness leads to some variants being, to use Darwin's words, naturally selected. The process going on within any one cage of our experiment is less easily describable as selection, rather, it is a form of 'screening' of one particular variant on the basis of its R_0 value. But this in no way detracts from the selective nature of the overall process which, it should be noted, would still prevail if the experiment was staggered; i.e. one mutant being tested for R_0 at a time. This form of selection, then, in which R_0 rather than w is the criterion of success or failure, may be described as *n*-selection. It should be noted that in *n*-selection the term fixation has no meaning; we talk instead of the *establishment* or otherwise of a mutant population.
>
> It should be readily apparent that *n*-selection is a much less rigorous form of screening than *w*-selection. If our gedanken experiment had involved mutant propagules sympatric with each other and with the wild type, rather than allo-

patric propagules, the outcome would, under normal laboratory conditions, have been fixation of the wild-type and extinction of all the other variants. Thus the different outcomes of *w*-selection and *n*-selection, and particularly *the power of n-selection to permit the establishment of a drastically altered morphotype with a lowered degree of coadaptation*, cannot be doubted. What most certainly can be doubted is whether conditions ever exist in nature that allow the fitness of a new variant to be assessed only in terms of R_0; that is, to be subjected only to *n*-selection.

There are three important differences between our gedanken experiment and nature. Basically, in nature:

1. Mutations of a particular kind usually occur singly, not in large groups.
2. Mutants appear in, and consequently compete with the other members of, a generally non-mutant population.
3. Mutants and non-mutants interbreed.

Taken together, these features of natural populations would appear to preclude the possibility that a natural macromutant would be subject only to *n*-selection. However, if we consider the matter further this impossibility becomes instead a very low but non-zero probability; and so, perhaps, we have a very rare (but very important) evolutionary event. I will take the three above points in turn.

3.5.1 Occurrence of Clusters of Mutant Progeny

There are at least three possible mechanisms for the production of such clusters: First, mutations in early germ-cell precursors. Sinnott *et al.* (1958) state: 'Mutations in cells ancestral to the gametes, such as spermatogonia, may affect several or many gametes, resulting in the appearance of clusters of mutant individuals in the progeny.' A second, more speculative possibility is the occurrence of a cluster of mutant individuals (not necessarily mutant *progeny*) through 'horizontal transfer of genetic material via RNA-based viruses' (Erwin and Valentine, 1984). Third, and most relevant to radical morphological changes, there is the possibility of ordinary germ-cell mutation in 'maternal effect' genes. Since the initial, key morphogenetic decisions in a wide variety of organisms are often made by the products of such genes, the clusters of mutant progeny produced by this method will often be morphological 'monsters'. Such a cluster can occur after a single generation if the mutation is dominant, but not until at least after two generations if the mutation is recessive.

3.5.2 Lack of Competition

While many gross morphological mutants are inviable or sterile, and so would fail even the relatively weak test of *n*-selection, others are perfectly viable, though vastly less fit than the wild-type when in direct competition with it. In nature, such mutants will not be physically isolated, as they were in our gedanken experiment. However, if a mutation is such as to alter their pattern of resource utilization so that they are no longer jointly limited, with the rest of their population, by the same set of limiting resources, then they will not be in competition with non-mutant individuals, despite sympatry. If neighbouring areas of ecospace are thinly populated, then they may not be in competition with individuals of other species either. It seems inevitable that some small fraction of morphological macromutations will achieve this sort of effect; and it is of interest that the ecological condition favouring them—a sparsely populated ecospace—is likely to have been much more prevalent in early evolution, when most major body plans arose, than it is today.

3.5.3 Interbreeding

If a cluster of macromutants characterized by a very divergent resource-utilization pattern comes into existence, it has two possible fates (apart from extinction) depending on its reproductive compatibility. If the mutants interbreed with wild-types, then we have a macromutational equivalent of Levene's (1953) multiple-niche polymorphism. If they are only self-compatible, then we have a new species characterized by (a) gross morphological difference from its congeners, including the parental species, (b) an ecological role in which its performance is poor but sufficient in the (ultimately temporary) situation of no competition and (c) a very low level of internal coadaptation resulting from a gross imbalance between major mutant allele and many, perhaps all, modifiers.

It is worth briefly considering the long-term future of a macromutant propagule which achieves spearate species status either by 'instantaneous speciation' as above, or by later operation of the Wallace effect (see Murray, 1972) in the Levene-type system, or by the appearance of the mutation in a population of asexual or hermaphroditic organisms rather than the cross-breeding population implicit in the above discussion.

Although the subsequent evolution of such a species cannot be predicted in detail, it would be dominated by a restoration of coadaptation by conventional Darwinian selection acting on variation in modifier genes downstream of the macromutation in the morphogenetic tree. The contrast between a 'morphological window' macromutation and one becoming established through n-selection is twofold. First, as we have seen, the former is established by conventional Darwinian selection, while the latter is not. Second, and of crucial importance in understanding the difference between these two postulated types of evolutionary event, the 'morphological window' macromutation sets up relatively little pressure for further change via modifiers because it is constrained not to disturb coadaptation much in the first place; whereas the 'monster' established through n-selection sets up an immense pressure for further change. Ironically, the non-Darwinian initial event sets the scene for a Darwinian revolution, whereas the Darwinian initial event does not. After the revolution, our major evolutionary transition is complete, and by then the initial n-selective change will comprise only a part of it.

It should be stressed, particularly as I have referred to the initial macromutants as monsters (to emphasize their lack of coadaptation), that these are *not* 'hopeful monsters' in the sense of Goldschmidt (1940). They are not well pre-adapted for their new ecological role; they are not the result of a progressive process of 'chromosomal repatterning'; and they are at most an *extremely rare* form of speciation. Thus in all these three respects—ecology, genetic basis and frequency—they differ from Goldschmidt's conception. (I consider the contrast further in Chapter 4).

Finally, it will be clear that n-selective establishment will not be possible for micromutants, because they cannot escape from competition or interbreeding with other members of their population. Probability of establishment, however low, is thus *positively* associated with the magnitude of morphological effect of mutations. This is the other reason (in addition to the one discussed in Section 3.4) why probability of evolutionary success is not a monotonically decreasing function of magnitude of morphological effect. Rather, the distribution takes the shape of a very asymmetrical U.

3.6 SUMMARY AND CONCLUSION

In this chapter, I outline four mechanisms for the divergence of body plans which are compatible with morphogenetic tree theory. I stress that, whether body plans are identified with higher taxa or with the higher levels of decision within a morphogenetic tree structure, there is not a clear division of levels (in a taxonomic or developmental hierarchy) into those which do, and those which do not, correspond to changes in body plan. Nevertheless, some morphological 'gaps' (e.g. between phyla) are obviously more major than others (e.g. between congeneric species), just as some levels of developmental decision are more major than others. The question arises as to whether the major differences in development that underlie the morphological differences among higher taxa have arisen merely by a compounding of many smaller changes over a long period of evolutionary time.

While strict neo-Darwinists claim that the answer to this question is a straightforward 'yes', and some anti-neo-Darwinists such as Goldschmidt (1940) give an equally categorical 'no', morphogenetic tree theory yields a less simplistic answer. Of the four mechanisms I describe, only primary divergence is exclusively micromutational. Secondary divergence is largely so, but may be permitted in the first place by a sudden change to a simpler morphology. Morphological windows and n-selection both operate through macromutational changes, though, as we have seen, the establishment of a new body plan can only be *initiated*, not completed, by an n-selective event.

It would be nice to be able to identify certain known events in morphological evolution with particular mechanisms, but this is rarely possible. For example, de Beer (1958) regards torsion as 'the chief problem of the evolution of gastropods', and seems, uncharacteristically, to suggest a macromutational origin for it (pp. 57–58). If there was such an origin, did it involve the crossing of a morphological window by a conventionally advantageous macromutation, or alternatively the establishment of an ill-coadapted 'monster' through n-selection coupled with long-term Darwinian refinement of the novel form?

Since I doubt that such questions can ever be conclusively answered, the only thing that can be done, in addition to stating what seem to be the *possible* mechanisms for major evolutionary transitions, is to make some comment on their relative frequency. I am convinced that, in pure 'frequency of occurrence' terms, primary and secondary divergence are orders of magnitude more common than evolution by advantageous macromutation (morphological windows), which in turn is considerably commoner than establishment of macromutants through n-selection. However, since the rarer processes cause greater morphological change when they do occur, the differences in 'amount of morphological evolutionary change' attributable to the different mechanisms would not be expected to be as great as the differences in their frequency of occurrence. It is for this reason that authors like myself, who propose only very rare macromutational change, are sometimes regarded as a threat to orthodoxy. The extent to which morphogenetic tree theory is indeed 'counter-orthodox' is considered further in the following chapter.

Chapter 4

Relationships with Other Evolutionary Theories

4.1 Introduction 50
4.2 Résumé of morphogenetic tree theory 51
4.3 Micro- and macromutational theories 53
4.4 Developmental evolutionary theories. 56
4.5 Palaeontological and systematic theories 59
4.6 Prospect. 63

4.1 INTRODUCTION

In the three preceding chapters I formulated various aspects of morphogenetic tree theory: the tree itself (Chapter 1), its general evolutionary patterns (Chapter 2) and the mode of origin of body plans suggested by those patterns (Chapter 3). The purpose of the present chapter is to discuss the relationship between morphogenetic tree theory as a whole and other main strands in recent and current thinking about evolution.

There are two important reasons for doing this. First, the significance of a new theory is only adequately appreciated if its interconnections with areas of existing theory are understood. In this context, both 'positive interconnections' and incompatibilities are of interest. Second, when a theory is first put forward, critics are apt to leap to unduly rapid conclusions about its relationships with pre-existing theories (as well as its inherent correctness or otherwise). For example, my own earlier presentation of morphogenetic tree theory (Arthur, 1984, Chapters 9–13) has been seen as correct and orthodox (Peel, 1986), correct but unorthodox (Cohen, 1985), wrong but orthodox (Forey, 1985), and wrong and unorthodox (Jones, 1984: this review contains numerous errors and misunderstandings, some of which are discussed in Chapter 5). I use 'orthodox', here, as meaning along broadly neo-Darwinian lines. Although it will never be possible to reduce to one a diversity of views on the extent to which a new theory is in keeping with the orthodoxy of the day, it should be possible to reduce its variance somewhat by pointing out what the interconnections of the new theory seem, to its architect, to be.

Since current evolutionary theory is in a state of healthy flux, the relationship between a new theory and neo-Darwinism is not the only thing that is of interest. This fact is reflected in my discussing *three* main interconnections in sections 4.3, 4.4 and 4.5 below. The first of these—interconnection between morphogenetic tree theory and theories stressing the importance of micromutation and macromutation—includes the relationship with neo-Darwinian views. Sections 4.4 and 4.5, however, are concerned with other relationships.

I make no attempt herein to cover existing evolutionary theories in an exhaustive way. In particular, I omit theories—important as they may be—that deal with a different evolutionary realm and whose fate will be largely independent of that of morphogenetic tree theory. An example of such a theory is, as pointed out at the beginning of the book, Kimura's (1968, 1983) 'neutral theory of molecular evolution'. Whether this turns out to be right or wrong, or whether, as Lewontin (1985) has suggested, one large class of polymorphisms is neutral, the other maintained by some form of balancing selection, does not affect the correctness or otherwise of morphogenetic tree theory. This is in complete contrast to, for example, 'strict' neo-Darwinism (as 'defined' in Chapter 3): if strict neo-Darwinism is correct, morphogenetic tree theory, or at least part of it, is wrong, and vice versa.

Finally, since it has taken three fairly long chapters to characterize morphogenetic tree theory, it seems sensible to give a brief résumé of its main features before attempting to describe its interrelationships. This résumé is provided in the following section.

4.2 RÉSUMÉ OF MORPHOGENETIC TREE THEORY

The following points (1–15) represent the main proposals embodied in morphogenetic tree theory as a whole. I have grouped them into three 'blocks' corresponding roughly to Chapters 1, 2 and 3. Strictly speaking, the first block is 'the theory of the morphogenetic tree', and this is a *developmental*, not an evolutionary, construct. The two subsequent blocks jointly comprise 'the theory of morphogenetic tree evolution'. The developmental and evolutionary components are distinct but have a definite relationship to each other: the evolutionary component is dependent on the developmental but not *vice versa*. It is useful to have a general term to cover both components (and hence the whole series of points below), and I use the vaguer term 'morphogenetic tree theory' for this purpose.

1. The overall developmental system of a multicellular organism, or at least the 'becoming different' aspect of it (see Chapter 1 for an explanation), is based on a *hierarchical, tree-like* arrangement of causal links.
2. Corresponding to this hierarchy of causal links is a hierarchy of loci (D-loci: Arthur, 1984).

3. The average magnitude of phenotypic effect of mutations at these loci decreases going 'up' the tree (i.e. with lateness of developmental activity of the loci concerned).
4. Each locus is represented not by a single magnitude of effect but rather by a *distribution*, indicating that there is a within-locus, as well as an among-locus, component to the variation in 'magnitude of phenotypic effect of mutations'.
5. These distributions are such that some of the earliest-acting loci are incapable of micromutation. (They *are* capable of zero-effect mutation: see Chapter 1.) *small change actually means big change early on!!*
6. Morphogenetic trees (as outlined in points 1–5) are capable of three types of evolution: *phase* change, *structural* change and *distortional* change (details in Chapter 2).
7. All of these three types of evolutionary change occur more frequently in later developmental stages than in earlier ones.
8. This is due to a systematic trend in constraint through developmental time, which is associated with the number of modifiers 'downstream' of each causal link in the tree structure (or downstream of each locus, in the genetic version).
9. When they do occur, evolutionary changes in early stages have more radical morphological effects than changes in later ones.
10. Phase change occurs more frequently than structural change. This is because of their genetic bases: new alleles and new loci, respectively. (I find it hard to estimate the relative frequency of distortional change because its genetic basis is less clear, but I would hazard a guess that its frequency is intermediate between those of the other two.)
11. At any moment in evolutionary time, a morphogenetic tree has a 'leading edge'—the terminal level of causal links in which most changes (whether phase, structural or distortional) are occurring.
12. Because of *structural* change, the distance of this leading edge from the base of the tree will alter as the tree grows or simplifies. (In the latter case, we should perhaps talk of a 'receding edge'!)

13. Alternative body plans can arise in any of the following ways:
    (a) Long-term micromutational divergence in the leading edges of trees in two different lineages, coupled with growth of the trees. This is *primary divergence* if its starting point is a primitively simple tree, *secondary divergence* if it starts from a tree with derived simplicity, achieved, for example, by progenesis.
    (b) Fixation of an advantageous macromutation affecting an early causal link, coupled with subsequent minor modifications of the macromutation's effect. In this case, one lineage has crossed through a *morphological window*. The origin of the Clausiliidae—a family of sinistral

snails—was probably of this sort.
  (c) Establishment of an ill-coadapted macromutant by n-selection (see Chapter 3), coupled with a major revolution in its 'downstream' modifiers.
14. With regard to relative commonness, it is probable that (a) > (b) > (c).
15. Statements 7 and 14 are essentially different ways—continuous and discrete—of looking at the same phenomenon; except that considerations of (i) the non-independence of characters and (ii) n-selection both suggest a slight increase in the frequency of evolutionary changes at the very top end of the spectrum of 'magnitude of effect'.

This, then, is morphogenetic tree theory in outline. We now enquire into its relationship with neo-Darwinism and macromutational theories (Section 4.3), theories of the 'developmental evolutionists' such as Waddington (Section 4.4), and an assortment of palaeontological and systematic theories (Section 4.5), including punctuated equilibrium and cladistics.

4.3 MICRO- AND MACROMUTATIONAL THEORIES

Throughout the history of evolutionary biology, there has been a recurring argument between those who see evolution as a slow, gradual process involving many tiny changes and those who see it as a process involving sudden, radical changes either as well as, or instead of, long-term accumulation of minor modifications. The debate began with the exchange between Darwin and Huxley, mentioned in the previous chapter; it continued with periodic attacks on the micromutational view by de Vries (1905), Goldschmidt (1940), Willis (1940), and many others.

I think it is now well established that the view of trans-specific evolution as being predominantly macromutational is wrong; and it is certainly not compatible with morphogenetic tree theory. However, it is worth briefly examining the best-developed macromutational view—Goldschmidt (1940)—so as to pinpoint what its key features are; and to contrast these features with the evolutionary process I have described in Chapters 2 and 3. The reason for doing this is that what Goldschmidt said is often misunderstood, and consequently the relationship between his work and that of others is erroneously stated.

Basically, Goldschmidt (1940) regarded intraspecific evolutionary changes as being entirely distinct from those involved in the origin of species and higher taxa. He states (p. 396): 'There is no such thing as incipient species. Species and the higher categories originate in single macroevolutionary steps as completely new genetic systems.' The contrast with neo-Darwinism is clear. For example, Mayr (1963) sets out the view that geographical races (particularly insular ones) are incipient species. He goes on to say (p. 589): 'Every species is an incipient new genus, every genus an incipient new family, and so forth.'

One problem with Goldschmidt's theory is that some of his terms, such as 'single macroevolutionary step' in the above quotation, have the effect of misleading his readers about what he actually meant. His 'single steps' are only single at the phenotypic level. They are produced by a *series* of genetic steps: a series collectively referred to by Goldschmidt as 'chromosomal repatterning' (p. 206). He envisaged a 'storage', in the genome, of individually detrimental mutations which were phenotypically silent until they collectively produced an integrated new phenotype. At this point they would suddenly be manifested—as a 'systemic mutation' (p. 334). Some resulting phenotypes would, by chance, be well pre-adapted to particular ecological niches. These were the 'hopeful monsters' (pp. 390–393). An obvious weakness in this scheme is the need for some mechanism to link phenotypic expression with selective value. Such a mechanism would be basically Lamarckian, though Goldschmidt did not see it as such, and we have no reason to believe in its existence.

A classic case of misinterpretation of Goldschmidt's work can be found in Riedl (1978, p. 167). Riedl takes Goldschmidt's 'single steps' at face value, rejects them, dismisses Goldschmidt as being wrong, and goes on to devise a 'storage hypothesis' of his own which is, ironically, a repetition of what Goldschmidt was really saying.

A final point about Goldschmidt's scheme is that he regarded his hopeful monsters as being sufficiently well integrated and preadapted to be acted upon favourably by *conventional* Darwinian selection. For example (Goldschmidt, 1940, p. 216): 'our systemic mutation... leads at once so far toward the new type that selection can immediately be efficaceous'. He gives as an example (p. 390) a mutant of *Archaeopteryx* with a fanlike arrangement of tail feathers which 'was a great *improvement* in the mechanics of flying' (my italics).

We can now recognize three major assertions of Goldschmidt's (1940) thesis:

1. Trans-specific evolution is *predominantly*, perhaps even *exclusively*, macromutational.
2. The macromutations are built up gradually at the genetic level but appear suddenly at the phenotypic level.
3. Macromutational variants, or at least the subset of them contributing to evolutionary change, are both well integrated and ecologically preadapted. They are selectively advantageous in the conventional Darwinian sense. They are 'monsters' only in the sense of being a large departure from the norm.

As pointed out in Chapter 3, my proposal for the occasional establishment of a 'monster' through n-selection differs in all three respects from Goldschmidt's scheme. The only point of connection between his scheme and morphogenetic tree theory is that in the case of 'morphological windows' (such as the gastropod chirality mutation) point 3 above applies.

Finally, I should state the obvious: variation among views on a particular biological topic has within- and among-person components! Goldschmidt's views evolved over time, and my statement of them is based on his major evolutionary work (1940). However, he makes clear in that work (p. 309) that his own earlier views were more moderate in relation to the frequency of macromutational change, and a later paper (Goldschmidt, 1952) seems also to be more moderate in this respect.

I have made it clear above that morphogenetic tree theory (a) is not, in general, a macromutational theory, since it involves *rare*, not predominant/exclusive macromutation; and (b) is not, more specifically, in line with the views of Goldschmidt (1940). Does that mean that it is a micromutational, neo-Darwinian theory? We began to address this issue in Chapter 3, and I pointed out then that morphogenetic tree theory is mutually exclusive with the strict form of neo-Darwinism but perfectly compatible with the broad form. However, I dismissed strict neo-Darwinism too quickly, and this is the appropriate place for a fuller discussion of this evolutionary view.

The danger here is that I set up a strict neo-Darwinian 'strawman' which is easily refuted, in the same way that, as Lewontin (1974) has pointed out, some 'selectionists' in the molecular evolution debate erect a neutralist strawman for easy refutation. Specifically, branding strict neo-Darwinism as an *exclusively* micromutational view, while in keeping with statements that some of the supporters of this view make, could be regarded as deliberately putting strict neo-Darwinism in a form which leads to its immediate rejection. After all, macromutations *do* contribute to evolution: I gave as examples in Chapter 3 the evolution of pigmentation in Lepidoptera and the evolution of chirality in gastropods.

There is a problem here. If we let strict neo-Darwinism take the more enlightened form of allowing occasional macromutation, does it not become equivalent to broad neo-Darwinism and morphogenetic tree theory? That is, are we not left with only one micromutational theory—a theory of *predominant* micromutations?

I think that the answer to these questions is 'no', and that there is still a real difference between strict neo-Darwinism and morphogenetic tree theory, as follows. The strict neo-Darwinian may acknowledge that macromutations occasionally contribute to evolution, but he sees no particular *pattern* in their occurrence and he sees them as events with no particular *importance*. In morphogenetic tree theory there is a very definite evolutionary pattern of occurrence—decreasing probability of contributing to evolution with increasing magnitude of effect (with a slight upward twist for mutations of maximal effect—see Chapter 3). In strict neo-Darwinism, the fall-off in this probability to the vicinity of zero is taken to be so rapid that there is no real pattern. With regard to importance, in morphogenetic tree theory the ability of some macromutations to open up new areas of morphospace for 'occupation' is an important concept; in strict neo-Darwinism it is not.

I end this section by repeating three key points from earlier chapters. First, as noted in Chapter 3, I see morphogenetic tree theory as an attempt both to visualize in more detail, and to explain, the pattern of predominant micromutation and occasional macromutation implicit in broad neo-Darwinism. Second, as pointed out in Chapter 1, morphogenetic tree theory belongs to the 'weak European school' identified by Gould and Lewontin (1979) in which morphological evolutionary patterns are seen as a composite of patterns in the occurrence of mutations on the one hand and patterns of ecological change and selection on the other. The patterns are emphatically *not* seen in exclusively ecological terms. Finally, it has unfortunately been necessary, in this chapter and the last, to resort to the arbitrarily defined terms 'micromutation' and 'macromutation', in order to make connections with the work of earlier authors who used these terms. However, as noted in Chapter 3, there is really a continuum of 'magnitude of effect', and it is preferable to think about this issue in terms of such a continuum. As Darwin (1859) said: 'Monstrosities cannot be separated by any clear line of distinction from mere variations'.

4.4 DEVELOPMENTAL EVOLUTIONARY THEORIES

Running parallel to the development of the modern synthesis from the 1930s to the 1950s, and to the refinements of it in the following two decades, was a progression of works approaching evolution from a developmental angle, most notably those of Waddington (1940, 1957), Schmalhausen (1949), Løvtrup (1974) and Riedl (1978). Unlike Goldschmidt's (1940) theory, whose relationship to the modern synthesis (direct antagonism) was clear, the ideas of the group I loosely refer to as the 'developmental evolutionists' were largely tangential to the synthesis, and were variously seen as 'for' it, 'against' it, or unconnected with it. The reason for this difference in the way the theories of Goldschmidt and the developmental evolutionists have been interpreted must, I think, lie largely in Goldschmidt's giving macromutation such a central role. The other group generally acknowledged that macromutations would make contributions to evolution, but did not place them in such a pivotal position.

Although Waddington, Schmalhausen, Løvtrup and Riedl formulated their ideas largely independently of each other, there are four central themes that pervade the work of this 'school' and give it some coherence. These are:

1. The claim that development is in some way hierarchical and that this has evolutionary significance.
2. The view that macromutations will contribute to evolution, albeit rarely, and will have important effects when they do so contribute.
3. A conviction that the 'staying the same' side of development (canalization etc: see Chapter 1) is evolutionarily relevant, though perhaps more to evolutionary stasis than to evolutionary change.
4. The idea that the environment interacts with developmental processes in a way that lets it play a more active evolutionary role than that of a 'sieve', as in the idea of genetic assimilation.

It will be readily apparent that morphogenetic tree theory is a continuation of the tradition represented by the first two of these central themes. I share conviction 3 but, as I pointed out in the first chapter, I cannot see, except rather vaguely, how canalization might be usefully pictured in a general way. I accept the idea of genetic assimilation, but do not assign to it any particular importance beyond acknowledging that it probably does sometimes happen outside the laboratory (and certainly inside it: see Waddington, 1953, 1956).

As in the previous section when discussing macromutational theories, I will concentrate here on the work of a single author in order to keep the discussion brief. Specifically, I will concentrate on the connection between morphogenetic tree theory and the work of Waddington, partly because I regard some of his ideas (in Waddington 1940) as, in a sense, the start of it all, and partly because his efforts can be said to be the most complete of the 'developmental evolutionist school', since he included all four of the central themes, and was instrumental in founding three of them (1,3,4). Anyone who wants a brief history of the relationship between the modern synthesis and developmental work in general should consult Hamburger (1980), who says of Waddington's (1957) 'Strategy of the Genes' that it could be considered as a draft of the missing embryological chapter of evolutionary theory: an opinion which I think is a little on the optimistic side, despite my sympathy for Waddington's approach.

The relationship between morphogenetic tree theory and Waddington's work is an interesting one, showing points of correspondence, points of disagreement and points of non-connection. Briefly, the relationship is as follows. Waddington (1940) proposed that the causal structure of development is hierarchical. He pictured a series of developmental decisions as a 'branching track system', and his diagrams of this have an obvious resemblance to the causal-link version of the morphogenetic tree.

Although there is a common grounding in hierarchy, there is also a fundamental difference between Waddington's approach, in his branching track system (and in his better-known epigenetic landscape: see below), and the morphogenetic tree approach. Waddington considered *one developmental process* confronted with a hierarchical pattern of decisions; in morphogenetic tree theory the emphasis is on the *whole developing organism* as a hierarchical causal structure. These two approaches are not incompatible, but they do represent very different emphases.

The epigenetic landscape was produced from the branching track system by adding to the model a component representing canalization. Although, as noted in Chapter 1, Waddington makes the comment that canalization corresponds to cross-linkages between different processes (and we can envisage it this way in morphogenetic tree theory as cross-links between branches), he did not actually build it into his model in this form. Rather, he pictured his branching tracks as a hierarchical arrangement of trajectories on an inclined plane, and built around them a 'landscape', using a third dimension projecting upwards from the plane, which placed the trajectories in valley bottoms. (The valleys *diverge*, in contrast to the convergence of real valleys.) The idea was that, given no disturbance, a developmental process would proceed through the valleys, making a particular decision at each bifurcation. Given a small disturbance, the process would be displaced slightly up the side of the valley, but would then 'flow' downhill and would resume its normal course. Waddington called the normal course the 'creode', and the tendency to return to it 'homeorhesis' or 'canalization'. There is an obvious analogy, as Waddington points out, with local stability around an equilibrium point, except that in Waddington's case the equilibrium is a process, not an end-state.

While the epigenetic landscape is generally recognized as a clever abstract picture of a complex developmental process, the 'landscape' aspect of it (representing canalization) has had a tendency to become predominant in subsequent thinking about this model, both Waddington's and other authors'. Canalization features very strongly in the evolutionary discussions of later chapters in *The Strategy of the Genes* (1957), while the hierarchical system of decisions around which the canalization was built up gets little attention. Indeed, Waddington (1957) does not give an index entry for 'hierarchy', or even for his own earlier 'branching track system'. Yet there is no doubt that this hierarchy is central. It is implicit in his diagram of the epigenetic landscape (1957, p. 29), and he states (p. 30): 'The number of separate valleys must increase as we pass down from the initial towards the final condition'. (Note that 'down' in an epigenetic landscape is equivalent to 'up' in a morphogenetic tree.)

The reason why Waddington did not go on to explore fully the evolutionary consequences of a hierarchical structure of development does not lie entirely in his being preoccupied with canalization. In addition, such an exploration must have been hindered by his view of the developmental role of the genes. Unlike the morphogenetic tree approach, in which hierarchy of causal links is taken to imply, *ceteris paribus*, hierarchy of development-controlling genes (D-genes), Waddington's approach did not extend hierarchical organization to the genic realm. Rather, he pictured the relationship between genes, on the one hand, and the branching track system/epigenetic landscape on the other as being very complicated, as indicated by his diagram of the epigenetic landscape 'from below' (1957, p. 36), and his earlier comment (1940, p. 83) that 'the course of each branch... is controlled... by the whole genotype or the greater part of it'. (Waddington is using 'genotype' here in the sense of the modern word genome.)

Since the idea of a hierarchical *genetic* causal structure, embodying early key decision-making genes, is closely associated with the idea of occasional macromutational inputs into evolution, it is not surprising that Waddington's lack of concentration on the hierarchical side of his own work was accompanied by a lack of explicit comment on the evolutionary role of macromutation. He was clearly not against such a role, though he correctly dismissed Goldschmidt's (1940) bizarre mechanism of chromosomal repatterning as a genetic basis for the origin of macromutant phenotypes. He states (1957, p. 115): 'there is no absolute necessity to suppose that the hopeful monster is brought into being by anything more strange than a normal gene mutation', and 'once it has appeared and begins to be favoured by selection, the remainder of the genotype will become reorganized around it'. But this is about as far as he goes.

In the end, most of Waddington's (1957) *evolutionary* ideas revolved around canalization rather than hierarchy, or were concerned with genetic assimilation, which has no real dependence on *either* of the two main features of his epigenetic landscape. Thus Waddington ended up concentrating on the third and fourth 'central themes' identified above, and said relatively little about the first two. Morphogenetic tree theory has precisely the opposite bias (a concentration on themes 1 and 2), and is thus largely complementary to Waddington's work.

I would like to make one final comment on Waddington. He violently objected (1943) to Mather's (1943b) equating of major genes with decision-making genes, and polygenes with the buffering genes responsible for canalization. This seems an eminently sensible objection. As we have seen (Chapter 1), the decision-making or 'becoming different' aspect of development is fundamentally distinct from the 'staying the same' aspect, comprising canalization, co-ordination and repeatability. Each 'sector' presumably has its major and minor processes, and genes controlling them. To conflate these two qualitatively different kinds of division of the genes contributing to development is very confusing and should be avoided.

4.5 PALAEONTOLOGICAL AND SYSTEMATIC THEORIES

In this section I will discuss the relationship of morphogenetic tree theory with the theories of punctuated equilibrium and species selection (Section 4.5.1); cladistics (Section 4.5.2); and extinction theory (Section 4.5.3). This is a diverse collection of theories, the only connection among them being that they have been proposed by palaeontologists and systematists rather than by population geneticists or developmentalists: hence this section's rather general title. I will keep the discussions of individual theories briefer than in Sections 4.3 and 4.4 because the links with morphogenetic tree theory are, I think, less strong. However, it would be remiss to omit any comment, especially on the relationship between an explicitly hierarchical theory of development and an explicitly hierarchical theory of natural classification (cladistics); and between two theories—morphogenetic tree theory and punctuated equilibrium—both of which deal specifically with the morphological realm.

4.5.1 Punctuated Equilibrium and Species Selection

Like many 'neontologists', I have been rather critical of the theory of punctuated equilibrium (Arthur, 1982a, 1984). However, most neontological criticisms, including my own, have been directed at the punctuations rather than the long periods (sometimes several million years) of stasis. While the criticisms remain, their significance is much reduced if we regard the main importance of the punctuated equilibrium pattern as the stasis rather than the punctuations, as urged by Williamson (1981), Gould (1982) and Gould and Eldredge (1986).

What is the relationship between long periods of stasis that are only occasionally 'broken out of', and morphogenetic tree theory? Initially, it might seem that there is no relationship at all, since, as pointed out in Chapter 3, punctuated equilibrium is a pattern in macroevolutionary time, whereas morphogenetic tree evolution occurs on a megaevolutionary timescale. Certainly, there is no correspondence between the occasional advantageous mutations 'far down' the morphogenetic tree and punctuations. The typical intrageneric punctuation involves morphological change that may be great in relation to typical intrapopulational variation, but punctuational changes are decidedly at the small end of the spectrum of 'magnitude' with which morphogenetic tree theory deals.

I no longer accept that a phase of morphological stasis lasting for, say, 2 MY, despite many marked ecological changes, can be explained solely in terms of stabilizing selection emanating from external, ecological sources. The idea that developmental constraint is somehow involved—advanced in the initial formulation of punctuated equilibrium (Eldredge and Gould, 1972) and reiterated many times since then (e.g. Williamson, 1981)—seems unavoidable. But we have to go beyond this and ask what *kind* of developmental constraint. Although it is not generally recognized, there are two rather different kinds of constraint, as follows. First, there is constraint in the form of canalization, which, as I have made clear from the outset, is not dealt with by morphogenetic tree theory in its present form though I suspect it may be based on some sort of cross-linkage between different 'branches'. Second, there is the constraint that *is* explicitly dealt with by morphogenetic tree theory, namely the constraint associated with the number of downstream modifiers. The latter kind of constraint is minimal at the late developmental stages that are presumably involved in the stasis-breaking punctuations, and may well be irrelevant to punctuated equilibrium. However, 'canalization constraint' is presumably strong at *all* developmental stages, and this would seem a better candidate for the agent underlying stasis.

This proposal seems almost to have been anticipated by Eldredge and Gould (1972) and Williamson (1981), who talk of developmental 'homeostasis'. Yet surprisingly none of these authors refer to Waddington (1957). 'Developmental homeostasis' is a limited, end-state-only way of looking at Waddington's homeorhesis (discussed in the previous section), which in turn, as we have seen, is equivalent to canalization. It seems to me that the explanation of stasis sought by the founders of punctuated equilibrium, and many of its supporters, is to be found in Waddington's 'canalization' albeit we do not yet understand how this process actually works.

If the above interpretation is correct, then punctuated equilibrium theory and morphogenetic tree theory have in common the central theme of developmental constraint; but they involve two fundamentally different kinds of constraint, which have effects on two different (though intergrading) timescales.

If morphological change is concentrated in speciation events, then of course long-term evolutionary directions must be largely achieved in terms of differential success of species. This is Stanley's (1975, 1979) 'species selection', operating both through differential rates of production (by speciation) and of disappearance (through extinction). While it cannot be denied that species selection happens, this concept is not very useful in relation to theories of the evolution of *morphology* as opposed to taxa. Morphological innovation begins as a *within*-species phenomenon, regardless of whether it can occur at any time in a species' history or is confined to initiating speciation events. Thus I see little useful connection between morphogenetic tree theory and species selection—in contrast to the potentially interesting connection with punctuated equilibrium.

As an aside, one neontological criticism of punctuated equilibrium that *does* concern stasis is that it is erroneous to suppose, as Williamson (1981) does, that 'The long-term morphological stasis noted by punctuationists in the fossil record is clearly mirrored by the relative morphological uniformity of most widely distributed modern species'. Of course, we could quibble about what Williamson means by 'relative', but modern species are emphatically *not* uniform, and often consist of well-marked geographical races, the differences between which are nearly as great as interspecific differences in the same genus. A case in point is the large-shelled race of the landsnail *Cepaea nemoralis* in western Ireland and the Pyrennees. We may be puzzled by the geographical distribution of this race, and we may have only limited data on its genetic basis (though we do have *some* such data: Cook 1967; Cook and Cain, 1980), but its existence cannot be disputed. Cases such as this point to a species being a series of parallel lines through time (in the context of a morphology-versus-time 'graph'), not just a single line. The problem, in general, is that neontologists see many populations at an instant in time, while in the few cases where palaeontologists can observe long-term history in detail it is usually the history of a single population (or small group of populations). What we really need are studies which are both spatially *and* temporally extensive; then we would have more data and, perhaps, less disagreement.

4.5.2 Cladistics

One of the main reasons for Hennig's (1966) formulation of the cladistic method (or phylogenetic systematics, as he termed it), was a concern for basing classification on genealogical, as opposed to morphological, aspects of evolution. That is, the emphasis was on the pattern of evolutionary relationships among species, not on the evolution of morphology and its developmental basis (see, for a brief introduction to the rationale, Hennig 1981, Chapter 1). Clearly, then, the approach of morphogenetic tree theory is very different from the original cladistic approach. In the transformed school of cladistics (see, e.g., Platnick, 1979), the emphasis on genealogy (and indeed on evolution in general) was dropped, and the concepts of homology and natural classification, denuded of their usual evolutionary interpretation, came to the fore. Transformed cladistics, then, is even further from morphogenetic tree theory than its Hennigian 'ancestor', because the difference is not just between two approaches to evolution—genealogical and morphological—but rather a difference between one theory which explicitly deals with evolution and another which does not.

Now this is not a promising background against which to develop any interconnections between the two areas of theory. Yet there are at least two such interconnections that deserve mention, one of them 'positive', the other 'negative'. On the positive side, there is a connection between the idea of nested homologies fundamental to the whole cladistic approach, and causal hierarchy of development, as incorporated in the morphogenetic tree approach. Consider, for example, two homologies, one nested within the other in a cladistic natural classification. (The same point could be made in relation to a more complex group of homologies; in general, the issue here is the 'polarity' of homologies: see Patterson, 1982.) I will take as an example the tetrapod pentadactyl limb and the primate forelimb/hand. The usual interpretation of two such homologies, one 'nested' inside the other in a natural classification, is that the 'outer' homology appeared first *in evolution*. In morphogenetic tree theory, characters constituting the outer homology also begin to appear earlier *in development* than those specific to the inner one. While this developmental interpretation is not unique to morphogenetic tree theory, nor the nested homology concept unique to cladistics, the centrality of the concept of hierarchy in these two theories—and the link between both of them and von Baer's law—makes the connection particularly clear.

The 'negative' connection between morphogenetic tree theory and cladistics concerns the question of how morphological changes actually come about, as distinct from the taxonomic distribution of the resultant characters at a particular moment in evolutionary time. The cladistic procedure is a generally 'equable' one in that it gives equal weighting to different cladogeneses, and to different characters (except inasmuch as they weight themselves: see Patterson, 1982), and refuses to recognize the concept of evolutionary 'grade'. Since cladistics is essentially a ranking procedure, producing 'cladograms', which are ordinal as opposed to interval constructs (see Siegel, 1956 for an explanation of these terms), it makes internal sense that it treats morphology in this equable way. However, there is a stark contrast with morphogenetic tree theory, where both development and evolution are seen to contain a variety of magnitudes of effect of morphological steps. Ironically, the equability of cladistics fits better with a strict neo-Darwinian evolutionary mechanism than with morphogenetic tree evolution. No doubt many cladists would dislike this conclusion, for example Rosen (1984) who claims that 'the ghost of neo-Darwinism... will not haunt evolutionary theory much longer'. Nevertheless, the conclusion seems inevitable. This is not to say that the 'equable' approach is necessarily a bad way to produce a natural classification: in this respect it seems to me to have both advantages and disadvantages. But it is as constraining as the strict neo-Darwinism with which I have linked it when it comes to thinking about the evolution of development.

4.5.3 Extinction Theory

Much recent palaeontological study has focused on extinctions, especially on the question of whether they have a periodicity (of around 26 MY: see Raup and Sepkoski, 1984). Extinction theory and morphogenetic tree theory are in a sense opposites in that the former deals (as does neo-Darwinian theory) with destruction, while the latter deals with 'creation', by giving a developmental pattern to the occurrence of mutations. However, despite this contrast, the two areas of theory have something in common, and there is a possible link between them. Their common feature is their grounding in the megaevolutionary timescale (or the 'third tier': Gould 1985)—a scale of tens to hundreds of millions of years. The possible link between them is this. The extensive clearing of the ecospace by a mass extinction is a potential starting point for developmental revolutions initiated by instances of n-selection. Such a process would not, of course, be dependent on whether extinctions are indeed periodic (which remains to be established: see Patterson and Smith, 1987); but it would be facilitated if the extinction events were taxonomically extensive, as is now known to be the case.

4.6 PROSPECT

Morphogenetic tree theory, as formulated in this book (and, in more rudimentary form, in three earlier publications: Arthur, 1982b, 1984, 1987a), is a hypothesis, or, more correctly, a series of linked hypotheses. Some of these are directly testable, while others are not; though in the latter case it should be possible at least to bring data to bear on the issues concerned indirectly. Morphogenetic tree theory is *not*, as some reviewers of *Mechanisms of Morphological Evolution* have implied, either well known to be true, or well known to be false. At this stage, it can neither be rejected as incorrect nor accepted as something that was implicit in the modern synthesis all along.

While the requisite testing has yet to begin (but see Chapter 5), I would like to make two brief anticipations of what may follow it. First, as regards the likelihood of falsification of the theory of the morphogenetic tree itself, upon which the evolutionary side of the theory rests, I am confident that a hierarchical causal structure of development will be vindicated. Such a causal structure seems by far the most readily compatible with the extensive observational data that we possess on the development of multicellular organisms; and it has almost universally been favoured both by the 'developmental evolutionists' (Chapter 3) and embryologists themselves (Chapter 1). That this hierarchy extends to loci, and that the locus-based version of the morphogenetic tree (which couples genetic hierarchy with a decreasing *average* magnitude of effect through developmental time) is broadly correct, I am almost equally confident, although this proposal is obviously a little more risky, as it makes additional assumptions.

The most risky developmental proposal of all those I have made herein is that the *distributions* of 'magnitude of effect' of mutations are such that at least some key early-acting D-genes are incapable of effective micromutation. While I continue to believe that this is so, I should stress that if this proposal turned out to be false, it would not necessitate an abandoning of my picture of morphogenetic tree evolution, with its variable magnitude of morphological effect of evolutionarily successful mutations, and a return to the confines of strict neo-Darwinism and the concept of the 'infinitely plastic organism'. If the mutational version of the morphogenetic tree turns out to be correct, then neo-Darwinism in its strictest form is logically impossible; if, on the other hand, effective micromutation of all early-acting genes can occur, strict neo-Darwinism becomes possible but not necessary. In such an eventuality, whether or not a particular macromutation will give rise to an evolutionary change depends on whether the selective conditions are appropriate for it to do so, and *not on whether there is a possible micromutational route to the same morphology*. After all, if a rare advantageous macromutation arises, and the ecological conditions for its spread are in existence, then it will simply spread, stochastic processes permitting. It does not stop to consider whether there is a less drastic way to achieve the same end. This illustrates the general point that, while the theory of morphogenetic tree evolution is ultimately dependent on the nature of the tree itself, this dependence is a little more subtle and incomplete than might initially be supposed.

Turning to a more positive 'prospect', in what directions should morphogenetic tree theory proceed in the future, if the corpus of it that I have advanced so far turns out to be broadly correct? Two such directions are clear at this stage. First, as I have stressed at various points, the idea of canalization needs to be incorporated, in an explicit way, into the morphogenetic tree picture. Second, we should progress from considering only the *magnitude* of morphological effect to considering also the *type* of effect. A hormonal change causing a slight increment to post-embryonic growth and a mutation affecting pattern formation in just one out of many embryonic fields at a late stage in embryogenesis are both 'small', yet they are very different from each other. Although I have taken a step in the direction of recognizing types of change in distinguishing phase change, structural change and distortional change, it should eventually be possible to achieve a more detailed classification of types than this. When we arrive at such a classification of types of mutational effect on morphology, we should be able to produce a theory of the evolution of development that builds on, but goes far beyond, the theory that I have advanced herein.

Chapter 5

Applying Morphogenetic Tree Theory

5.1 Introduction 65
5.2 General evolutionary trends 66
5.3 Origin of higher taxa 73
5.4 Conclusions 83

5.1 INTRODUCTION

So far, I have largely been concerned with formulating morphogenetic tree theory and with elucidating its relationships with other bodies of theory in evolutionary biology. The time has now come to *apply* the theory, that is, to examine a number of 'case studies' and to assess the extent to which they can be usefully interpreted in terms of morphogenetic tree evolution.

The issues to which morphogenetic tree theory is potentially applicable are inevitably large ones, and the question arises as to what exactly should constitute a 'case study'. There would appear to be two possible approaches. First, we could have the kind of case study wherein the genes controlling a particular developmental process, for example pattern formation in insects, are examined and their relative rates of evolution compared with a view to confirming (or rejecting) the hypothesis of more rapid evolution of genes active later in the process. Second, we could have a case study comprising an attempt to elucidate the origin of a particular 'higher taxon', e.g. Diptera, in a morphogenetic tree context.

While I do indeed adopt these two different approaches in Sections 5.2 and 5.3, respectively, and while they clearly have separate connections with earlier chapters (Chapters 2 and 3, respectively), it is important to stress that they are not as distinct as they might at first seem, for the following reason. Ideally, we would like to be able to identify the precise mutational events involved in the origin of a particular body plan such as that of the Diptera, but in practice it is doubtful if this will ever be possible. Given that unfortunate state of affairs, we can only talk in general terms about the broad *kind* of mutational events involved, in particular their time of earliest effect in the developmental process, and, associated with that, their average magnitude of effect. But considerations of mutational change at a particular time in development are most meaningful in a comparative context. For example, the changes involved in the production of the higher taxa, such as classes and orders, might be expected to originate earlier in development than those involved in the production of lower ones such as genera and species as I hope to show in Section 5.3. Since the proliferation of genera and species within, for example, an order, is a much more frequent event than the origin of novel orders, we have connected the two approaches. We started by considering the 'origin of a body plan' type of case study and have ended up with the other type of study, namely 'relative evolutionary rates of early- and late-acting genes'.

Having made the point that the two approaches are not distinct, why treat them separately in the following two sections? There are at least two answers to this question. First, it is often helpful, in general terms, to address a problem from a number of different angles. Second, and more specifically, while the two issues—relative evolutionary rates of genes active at different stages and origins of taxa at different levels—are clearly related, there is *not* a direct one-to-one correspondence between them, as I pointed out in Chapter 3. That is, we cannot equate evolutionary change of an early-acting gene with origin of a new higher taxon. I support the view that there is a general, probabilistic connection between the two, but not all evolutionary events involving early genes lead to new higher groups, and few if any of such groups have originated through evolutionary changes at one or a few early-acting loci without additional pervasive changes in modifier loci acting at later stages. So there is enough difference between our two approaches to pursue them independently, but enough overlap in the issues involved to warrant a continual attempt to elucidate the connection between them. Indeed, if we can eventually reach a consensus on the nature of the connection, that will be a considerable advance for evolutionary theory in general.

5.2 GENERAL EVOLUTIONARY TRENDS

I will restrict the discussion here to phase-change evolution, because, of the three categories of change among which I distinguished in Chapter 2 (phase, structural and distortional), this is the one upon which quantitative data can most easily be brought to bear. So the approach here will be to compare the evolutionary rates of different developmental stages, or the genes controlling them, and to examine the extent to which data on this topic are in keeping with the predictions of morphogenetic tree theory. I will not deal with the addition of new developmental stages or with situations in which some developmental processes are being shifted, in an evolutionary sense, between stages.

With regard to general trends in phase-change evolution, morphogenetic tree theory has one main prediction to make: more rapid evolutionary rates at later developmental stages. However, this apparently simple prediction requires some qualification, especially as it can be applied either to developmental stages themselves or to the genes controlling them. In the former case, both higher per-stage mutation rates and higher probabilities of mutations being selectively advantageous contribute to the prediction of higher allele-substitution rates at later stages, but smaller average morphological effects per substituted allele for genes controlling later stages produce a question-mark over whether *phenotypic* evolutionary rates will have the same bias towards late stages as allele-substitution rates. In the latter case—i.e. comparison of *genes* at different stages rather than the stages themselves—the prediction is simpler in two senses. First, it has only one cause, the difference in average probability of mutations being selectively advantageous at loci acting at different developmental stages. Second, since we measure genic evolutionary rates in terms of DNA sequence divergence and not in phenotypic terms, gene-based studies evade the problem of the 'counteracting' force mentioned above.

Because of the relative simplicity of morphogenetic tree theory's prediction about evolutionary rates at the genic level, and because of the recent accumulation of much data on genic divergence, I will concentrate, in this section, on comparing theory and data at this level of observation (5.2.2). However, I will first make a few brief comments (in 5.2.1) on comparison of theory and data at the 'higher' level of developmental stages.

5.2.1 Developmental Stages

Providing that the greater morphological change per substituted allele in the case of loci with early developmental effects does not outweigh the relative rarity of substitutions at such loci, the prediction of morphogenetic tree theory is that later stages will evolve more rapidly than earlier ones. A major result of this direction of bias in rates is that in cross-taxon morphological comparisons, later stages should show a greater difference than earlier ones. Now this, of course, is von Baer's law (his third law to be precise) which is widely accepted among evolutionary biologists and systematists (see, for example, Bonner 1974; Gould, 1977b; Patterson, 1982), and we thus appear to have a case of compatibility between the predictions of morphogenetic tree theory and the data. However, there are three problems here. First, the 'data' do not take a quantitative form and they permit different interpretations, though the sort of 'visual data' presented in Figure 5.1 to illustrate von Baer's law is quite persuasive. Second, the 'proviso' with which this paragraph started weakens the prediction of greater evolutionary rates at later stages. Finally, von Baer's law is not *entirely* accepted in the biological community, so a brief examination of objections to it is in order.

I have only heard two objections to von Baer's law, and one of them need not be taken seriously. This is the 'objection' that there are some cases in which the law is violated. Since most biological laws are of a probabilistic nature, and since *all* of them (even Mendel's!) have exceptions, such an objection is of no consequence unless the cases of violation are so numerous that the law does not

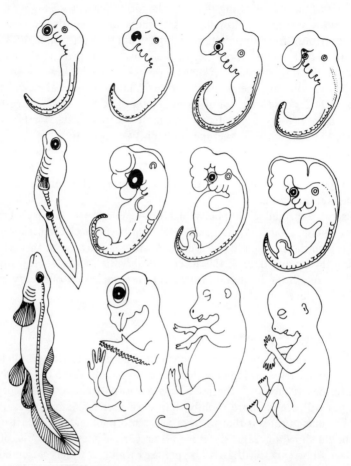

Figure 5.1 Embryos of (left to right) fish, hen, cow and human, showing early similarity (top) giving way to later differences (bottom). (From Raff and Kaufman, 1983)

even hold as a *general* one, let alone a universal one. There is little doubt that exceptions to von Baer's law are too few to constitute that kind of threat; and, as I pointed out in Chapter 2, the number of exceptions actually decreases if we consider the law to apply separately to the different morphogenetic trees of a complex life cycle.

The more serious, and more interesting 'objection' to von Baer's law is that it applies only after a certain stage in development and that there is a very early phase—up to or a little beyond gastrulation—where inter-taxon differences can be much more pronounced than the law would lead us to expect. If this is indeed the case, then the question arises as to why these very early stages are able to escape the high degree of constraint that apparently applies to those stages that immediately follow them. There is an approximate correspondence between the early 'evolutionarily flexible' stages and the control of development by maternal gene products rather than by zygotic genes, the switch between which occurs, in a variety of organisms, at about the beginning of gastrulation. However, it is difficult to see how this is relevant. A gene controlling a key early developmental process should be highly constrained evolutionarily regardless of whether it acts maternally or zygotically. So the correspondence between early flexibility and maternal control may be merely coincidental. Another possibility is that the very earliest stages do *not* embody 'key developmental processes' in that whatever follows them is in some way independent of them, unlike the tightly dependent inter-stage relationship that occurs later on. This is more likely to be true in 'regulative' than 'mosaic' types of embryo (which represent opposite ends of a spectrum rather than discretely different types). Thus we might expect von Baer's law to be applicable further 'back' towards the beginning of development, the more mosaic the embryo concerned. Perhaps comparative embryologists will be able to tell us whether or not this is the case.

To conclude: much of the 'data' of comparative embryology is summarized by von Baer's law, and the prediction of morphogenetic tree theory that later stages should evolve faster than earlier ones is seemingly compatible with this law. However, the prediction suffers from the 'proviso' relating to the magnitude of morphological effect per substituted allele mentioned earlier, and the 'data' suffer from being descriptive rather than quantitative. We now turn to an examination of *genes*, where both problems disappear, although, as we will see, others arise.

5.2.2 Developmental Genes (D-genes)

A considerable body of data has been built up over the past two decades on the rates of evolution of particular genes, as estimated from the degree of DNA sequence difference between any two extant species and the known or presumed point in geological time at which they diverged from their most recent common ancestor. Much of this data is not directly relevant to the issue being discussed here because the genes concerned are not D-genes in the sense of Arthur (1984, Chapter 10), i.e. genes 'controlling' some aspect of the developmental process. For example, the cytochrome genes, whose evolution has been much studied, are concerned with basic metabolic processes common to most cells. The haemoglobin genes, on the other hand, have a highly tissue-specific pattern of activity, yet they are still not D-genes because their activation in some places and inactivation in others is an end-result, rather than a cause, of developmental processes.

The most detailed information on DNA sequences of D-genes derives from studies of genes controlling segmentation and related processes in *Drosophila*. This group of genes includes both those specifying segment number (see Nüsslein-Volhard and Wieschaus, 1980) and those, such as the bithorax complex (BX-C, see Lewis, 1978), which are involved in the specification of segmental identities. One of the most striking discoveries about these genes at the

DNA level is that several of them contain a highly conserved region of 180 bp in length called the homeobox (McGinnis *et al.*, 1984a; Scott and Weiner, 1984; see also review by Gehring and Hiromi, 1986). Several authors have observed that the degree of conservation—first noticed among BX-C, the antennapedia complex (ANT-C) and *fushi tarazu*—is usually higher at the level of presumed protein primary structure than at the level of DNA base sequence (e.g. Carrasco *et al.*, 1984; Shepherd *et al.*, 1984), the difference being attributable to 'silent' substitutions in the third positions of many triplet codes. This of course suggests that the homeobox is translated (as is now known: see below), and that natural selection is acting to conserve the amino-acid sequence of the resultant 60-amino-acid protein. It also seems that this 'protein' is not an independent molecule, but rather a part of a larger one. The term homeodomain is thus used to refer to that part of the larger protein that is coded for by the homeobox.

The function of the homeodomain is not yet known in detail, but two general points have been noted. First, its apparent confinement to the products of early-acting D-genes suggests that it is in some way involved in embryogenesis. Second, its similarity to proteins known to have a DNA-binding function suggests that it too may have such a function (Laughon and Scott, 1984; Desplan *et al.*, 1985). These two suggestions as to the function of the homeodomain are of course compatible: one obvious way for a protein with a key developmental role to 'work' is to switch genes on or off, which would in turn be facilitated by possession of a DNA-binding region. Recent work has confirmed that the homeobox does indeed have a DNA-binding role (see Robertson, 1987).

So far, the only sequence conservation that I have mentioned is between loci *within* one species (*D.melanogaster*). This in itself is of evolutionary interest since it suggests the origin of some of the genes containing the homeobox from others (and ultimately from one ancestral gene) by gene duplication. However, the homeobox is also found in highly conserved form in a wide variety of species across several animal phyla. Both the nature of the taxonomic distribution of the homeobox and the levels of conservation are of interest.

McGinnis (1985) screened a number of species from a variety of phyla for homeoboxes of the *Drosophila* antennapedia type. His results, together with those of McGinnis *et al.* (1984b), may be summarized as follows. Homeoboxes are found in other insects as well as in crustaceans, annelids, echinoderms, cephalochordates, urochordates and vertebrates. They are absent in bacteria, fungi, plants, coelenterates, gastropods and cephalopods. As McGinnis (1985) points out, these results suggest that the homeobox originated before the protostome/deuterostome split and that it has been secondarily lost in molluscs, although the latter conclusion is no longer valid, since Holland and Hogan (1986) found that molluscs *do* contain homeoboxes. These authors also note that the homeobox is not associated with any particular developmental strategy, despite earlier claims of a link with metamerism.

With regard to the degree of sequence similarity between homeoboxes of different species, this is extremely high given the large taxonomic distances across which comparisons have been made. The most striking example of conservation is the discovery of an (unidentified) gene in humans whose protein product has only one amino-acid out of 60 different from the antennapedia homeodomain (98% sequence homology; Boncinelli *et al.*, 1985). Although this is unusually high, sequence similarities in the range 60–90% are common.

Given distantly related species with genes containing homeoboxes very similar to the antennapedia homeobox of *D.melanogaster*, the question arises as to what the homeobox-containing genes in these other species do. In particular, are they, like their *Drosophila* counterparts, genes with an important controlling influence in early development? It would appear that the answer, at least in some cases, is 'yes'.

One case in which it appears that homeobox-containing genes have a role in early development, like the *Drosophila* homeotics, is in the frog *Xenopus laevis*. Two such genes have been identified in *X. laevis*, and in one case (Carrasco *et al.*, 1984) the gene's transcripts are found predominantly during neurulation. In the other (Muller *et al.*, 1984) the transcripts are abundant in oocytes, though they are also found, at a lower level, in later developmental stages. Of course, the fact that the genes are active early in development does not mean that they *necessarily* control developmental processes, as Muller *et al.*, (1984) point out. This is similar to my earlier point about tissue-specific expression in relation to haemoglobin genes. Both features—early transcription and tissue specificity—should be found in the genes that *cause* developmental processes to take a particular form (i.e. D-genes), but also in some of the genes which merely respond to, and are the results of, those same developmental processes. In the case of *X.laevis* homeobox-containing genes, it may be that future work will show that they are indeed D-genes, and I would hazard a guess that it will. Also, some recent work on mice suggests an early developmental role for other vertebrate homeoboxes (Gaunt *et al.*, 1986; see also McGinnis *et al.*, 1984c).

If we assume that the homeobox-containing genes will in general turn out to be early-acting D-genes, then it is clear that such genes—or at least parts of them—are highly conserved by natural selection. Evolutionary changes clearly do occur in them, even in the homeobox regions themselves, but at a very low rate.

This finding would seem to fit in with the predictions of morphogenetic tree theory (see Chapter 2). However, we have to be a little careful about drawing such a conclusion. The predictions of morphogenetic tree theory on the evolutionary rates of early- and late-acting D-genes are *relative*. There is no prediction of any absolute rate of evolution for any specified set of D-genes. So the finding of 'slow' evolution of homeoboxes is only meaningful if late-acting D-genes, such as the polygenes, evolve more rapidly.

It is in trying to make this comparison that we immediately come up against the problem that, precisely because of their slight effects, polygenes cannot usually be identified, and consequently cannot yet be sequenced. The possibility therefore remains open that they include highly conserved sequences evolving as slowly as homeoboxes. We can at least rule out the possibility that the polygenes actually contain homeoboxes rather than some alternative but equally conserved sequence, because the number of copies of the homeobox per genome is quite low (Levine *et al.*, 1984; Carrasco *et al.*, 1984). But that is as far as we can go. However, there is at least a clear prediction from morphogenetic tree theory that when D-genes with later and smaller effects are eventually sequenced, they will be much less conserved than the early-acting ones to which the data we currently have relates. In time, this prediction will no doubt be tested, and unlike many other 'predictions' in other areas of evolutionary theory, it will clearly have been made *before* the accumulation of the data which test it!

It is necessary at this point for me to respond to some rather odd comments made by Jones (1984) in reviewing my earlier presentation of morphogenetic tree theory (Arthur, 1984). Jones notes the discovery of the homeobox in genes controlling early dependent in *Drosophila* and its high degree of conservation across disparate taxa. He goes on: 'These genes regulating development therefore appear to be more—rather than less—resistant to the accumulation of mutations than are many of those which code for structural proteins; an observation which contrasts with the prediction of theories which emphasize the importance of developmental mutants in the origin of major groups.'

The first problem with Jones's statement is that he makes the wrong comparison. Morphogenetic tree theory, as we have seen, makes a prediction for the relative evolutionary rates of early- and late-acting D-genes. It makes no prediction whatever on the relative rates of evolution of all D-genes *versus* all genes without a developmental role. However, we can induce the right comparison in Jones's statement by reading 'early-acting D-genes' in place of 'genes regulating development', and 'late-acting D-genes' for genes 'which code for structural proteins'; and it might be argued that this is what he was really trying to say.

Now, however, we come across a second problem. The greater resistance of early-acting D-genes to evolutionary change is precisely what morphogenetic tree theory predicts—not the opposite, as Jones seems to think. The problem here is that Jones confuses frequency with importance. One of the main proposals in morphogenetic tree theory is that mutational changes in key early-acting D-genes will only very rarely be turned into evolutionary changes, because they are usually highly detrimental, but on the rare occasions on which they do contribute to evolution they may be involved in the origin of novel body plans.

Finally, the idea that in general 'theories which emphasize the importance of developmental mutants in the origin of major groups' are misguided seems particularly odd. Since the major groups are largely delimited on morphological and embryological grounds, their origin *must* have involved 'developmental mutants'. This is not a theory, it is a logical inevitability.

The important issue, in relation to the establishment of major groups, is what *kinds* of developmental mutants are involved. Unfortunately, studies of the degree of sequence conservation in early-acting D-genes cannot shed much light on this, especially when we do not know in detail what the genes concerned do. For example, does one of the rare non-silent base changes in a homeobox that becomes incorporated into evolution do so because it creates just as radical a developmental alteration as some alternative base change which is not found to have been incorporated into evolution? (In this case the difference is between a rare advantageous major change and the typical disadvantageous major change.) Alternatively, does the rare base change that becomes incorporated do so because it has a particularly small developmental effect, in the extreme case being selectively neutral because it does not affect the function of the homeodomain?

We cannot yet answer such questions, and in general we have to admit to a high degree of ignorance about the origin of major body plans and the high-level groups of organisms which they characterize. Nevertheless, morphogenetic tree theory makes some rather general predictions about the origin of body plans (see Chapter 3), and it will be useful to examine these in the context of a particular example. I do this in the following section.

5.3 ORIGIN OF HIGHER TAXA

The development of ideas on the evolutionary origin of major body plans has been severely hindered by the polarization of views into micro- and macromutational camps. There are three main difficulties that are associated with this polarization. First, these two 'types' of mutation are really just opposite ends of a continuum of 'magnitude of effect', as I have repeatedly stressed. Second, the dominance of the micromutational view (at least over the last half-century) has meant that the origin of major body plans has been seen as a non-issue because it is taken to be merely a compounding, over long periods of evolutionary time, of a host of 'ordinary' microevolutionary events. Finally, information on how 'macro' a particular mutation contributing to the separation of two major groups was at the time of its occurrence is usually impossible to extract with certainty from any cross-taxon comparison involving extant groups, which makes resolution of the polarized views very difficult. There are a few exceptions to this loss of information, such as the origin of a family of sinistral gastropods from a dextral ancestor, where we can see that a large phenotypic change must have occurred all at once at some point in the evolutionary history of the group concerned. However, this is an atypical situation, and usually we are faced with the problem that a cross-taxon comparison at a high level—say between two classes or phyla—reveals a gulf in morphospace of such width and complexity that the possible routes of evolutionary divergence are very numerous, and include both purely micromutational routes as well as those involving occasional but important input from macromutations.

I am going to use the present section to suggest an approach which may get us out of this impasse, and to illustrate this approach with a 'case study' involving the origin of insect groups at different levels in the taxonomic hierarchy. The shift of emphasis that I want to make is from magnitude of effect on adult morphology to time of effect in development; and it is necessary at this point to consider (a) the relationship between these two variables, and (b) the advantage

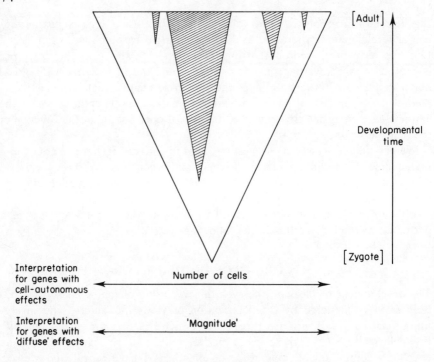

Figure 5.2 Diagram of proposed typical relationship between time of a D-gene's onset of activity and magnitude of effect of mutations in it. Outer (white) triangle represents overall development; smaller (shaded) triangles represent altered 'segments' of development due to mutations. Note different interpretations of x axis for different kinds of D-genes (see text)

of basing an approach to the origin of major groups on the latter. These two points are now considered in turn.

Figure 5.2 shows a proposed 'typical' relationship between how early a mutation begins to affect development and how great its effect is on adult morphology. This diagram is most easily interpreted in relation to mutations in genes with cell-autonomous effects, where the altered part of the adult is simply the clone of cells derived from the early embryonic cell in which the mutated gene is first switched on. However, even that statement requires some qualification. First, mutation in somatic cells is *not* being invoked here, despite the superficial similarity of Figure 5.2 to the pattern of mutated cell clones described by Wieschaus and Gehring (1976) which *is* based on somatic changes in the genetic material. Second, since the mutated gene may be switched off again in some sub-clone within the overall potentially-affected clone, it is always possible that the magnitude of morphological effect in the adult is less than 'expected'. Third, two or more clones may be altered, since a gene with a particular developmental effect may be switched on simultaneously in different parts of the body: polydactyly in man is an obvious example.

In the case of genes which exert a 'diffuse' effect on the phenotype, such as those which determine the production of a diffusible or otherwise transportable agent which can affect developmental processes 'at a distance', the interpretation of Figure 5.2 is more difficult. The generally increasing magnitude of effect on the adult with increasing earliness of action is still to be expected on the basis that early changes will tend to 'drive' later ones. However, it is much less clear than in the case of genes with cell-autonomous effects how the magnitude of effect should be measured, and this is reflected in the terms used to describe the x-axis in the figure for the two different types of gene.

It should be stressed that Figure 5.2 does indeed represent a proposed 'typical' relationship between magnitude and timing. No doubt there are examples of mutations which do not fit this 'triangular' pattern. For example, we can imagine a diamond-shaped area of the figure, representing a clone of mutant cells which dies out before the organism concerned reaches adulthood. However, a general, rather than universal, pattern will suffice for my purposes here.

Assuming that there is a general, probabilistic connection between macromutations and mutations with early effects, as shown diagrammatically in Figure 5.2, what is the advantage of focusing attention on timing rather than magnitude, given that, in a sense, they are two ways of looking at the same thing? In fact, there are two advantages, as follows. First, the terms 'early-acting gene' and 'late-acting gene' do not have the emotive power to distract attention from the continuum of effects that actually exists, whereas the terms micromutation and macromutation clearly do. Second, and much more importantly, while several compounded micromutational effects on adult morphology may be indistinguishable, after the event, from the effect of a single macromutation, several mutations acting to change late developmental stages *are* distinguishable from a single early change. Therefore, information on alternative evolutionary routes to two existing body plans separated by a large gulf in morphospace is *not* entirely lost, and cross-taxon comparisons of existing organisms may be useful even in the absence of any data on the development of extinct common ancestors.

An obvious question that now arises is this: is there a tendency for groups of high rank in the taxonomic hierarchy—such as classes or phyla—to diverge earlier in their development than lower groups? If the answer to this question is 'yes', that would seem to imply that the origin of the body plan characterizing the class or phylum is *not* simply a result of multiple occurrences of typical microevolutionary events. I will now proceed to examine this question in the context of a case study drawn from the development, evolution and taxonomy of insects. The following discussion has some common ground with the work of Rasmussen (1987), despite the use of quite different terminology and approach.

What follows is a three-point comparison involving groups whose relative positions in the taxonomic hierarchy and relative times of origin are known with certainty. The central question to be addressed is whether developmental pathways diverge earlier in comparisons across higher taxa than in comparisons across lower ones. The three 'points' of the comparison are:

1. The origin of the Insecta, involving divergence from its sister-group, the Myriapoda.
2. The origin of the Diptera, involving divergence from a sister-group which was probably the stem-group of the Mecoptera + Siphonaptera.
3. The origin of *Drosophila melanogaster*, involving divergence from the closely-related species *D. simulans* and *D. mauritiana*.

With regard to taxonomic 'levels', the group Insecta is now generally regarded as a subphylum (of the phylum Uniramia, which includes the Myriapoda: see Barnes, 1984). There is some disagreement as to whether Insecta = Hexapoda or whether the former is a subgroup of the latter. Barnes (1984) and Boudreaux (1979) equate the two groups, while Kristensen (1981; see Figure 5.3) includes one within the other. I will consider them to be synonymous here, though which approach is taken does not significantly affect the comparison I wish to make.

The Diptera is recognized as an order within the subphylum Insecta. I will assume that its sister-group is the Mecoptera/Siphonaptera stem-group, as depicted in Figure 5.3, and that this stem-group had characteristics close to the present-day Mecoptera (scorpion flies), including 'two pairs of similar, membraneous wings' (Barnes, 1984). That is, I am assuming that the wingless and laterally-compressed state of the Siphonaptera (fleas) is a recent specialization not found in the Mc/Si stem-group.

Drosophila melanogaster is of course a dipteran species, and its immediate evolutionary relationships may be summarized thus:

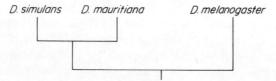

That is, its sister-group is the stem-species leading to the pair of species *D.simulans* + *D.mauritiana* (see Coyne, 1983). This split is not shown in Figure 5.3, but its location obviously would be at the extreme top right of the diagram.

Associated with the clear ranking of the three taxa (Insecta, Diptera, *D.melanogaster*) is an equally clear temporal ranking of the group origins. Perusal of Hennig's (1981) account of insect fossils suggests a minimum age of about 250 MYBP for the Diptera and 400 MYBP for the Insecta. Whether these minimum estimates are approximately correct, or whether the true ages are considerably greater, it is somewhat unlikely, to say the least, that their currently assigned temporal ranking is wrong. And since the split between *Drosophila melanogaster* and its sister-group almost certainly occurred within the last 10 MY, we have here as clear a temporal ordering of the three points in our comparison as it is possible to obtain.

I now want to focus on the main morphological differences between sister-

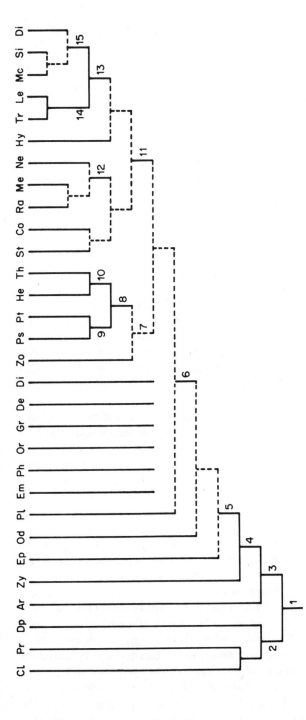

Figure 5.3 Proposed phylogeny of the extant hexapod orders. Monophylies which are considered well established are indicated by solid vertical lines. Higher categories denoted by numerals are: 1. Hexapoda, 2. Entognatha, 3. Insecta, 4. Dicondylia, 5. Pterygota, 6. Neoptera, 7. Paraneoptera, 8. Acercaria, 9. Psocodea, 10. Condylognatha, 11. Holometabola, 12. Neuropterida, 13. Panorpida, 14. Amphiesmenoptera, 15. Antliophora. Abbreviated ordinal names: Cl=Collembola, Pr=Protura, Dp=Diplura, Ar=Archaeognatha, Zy=Zygentoma, Ep=Ephemeroptera, Od=Odonata, Pl=Plecoptera, Em=Embioptera, Ph=Phasmida, Or=Orthoptera, Gr=Grylloblattaria, De=Dermaptera, Di=Dictyoptera, Zo=Zoraptera, Ps=Psocoptera, Pt=Phthiraptera, He=Hemiptera, Th=Thysanoptera, St=Strepsiptera, Co=Coleoptera, Ra=Raphidoptera, Me=Megaloptera, Ne=Neuroptera, Hy=Hymenoptera, Tr=Trichoptera, Le=Lepidoptera, Mc=Mecoptera, Si=Siphonaptera, Di=Diptera. (From Kristensen, 1981)

groups at each of the three levels, and in particular on their developmental origins, starting with *Drosophila*. According to Coyne (1983), a comparison of the morphology of the three species with which we are concerned here reveals that 'their only consistent morphological difference is the shape of the posterior process of the male genital arch (ninth tergite)'. No interspecific differences are known at the egg, larval or pupal stages. Thus we have here a morphological difference originating rather late in ontogeny and, as Coyne's (1983) study shows, being based on a polygenic system.

The most prominent morphological difference between the Diptera and its (presumed) sister-group is that between two very similar pairs of wings arising from meso- and meta-thoracic segments, and the reduction of the metathoracic wings to halteres in the Diptera. This is clearly an adult difference whose embryological beginnings precede those of the genital arch differences in *Drosophila*, and extend back in time into the larval stage, albeit being restricted to the larva's imaginal discs. (This conclusion is based on the fact that the BX-C mutations, which 'mimic' this evolutionary change, affect the discs.) However, overt larval differences also exist between the Diptera and its sister-group. Again assuming that the Mc/Si stem-group had characteristics closer to the present-day Mecoptera, with the Siphonaptera being more recently 'specialized', then the stem-group would not have had legless, dipteran-type larval forms ('maggots'), but rather 'caterpillar-like larvae with three pairs of short legs', as Barnes (1984) describes the typical situation in the Mecoptera. In conclusion, it is clear that differences between the Diptera and its sister-group extend much further back into development than differences between *Drosophila melanogaster* and its sister-group.

When we turn to the Insecta/Myriapoda comparison, we encounter a number of difficulties stemming both from lack of developmental information and lack of taxonomic agreement. In particular, our knowledge of myriapod embryology is much more limited than for insects; and there is some uncertainty over which high-level groups within the Arthropoda are monophyletic, and, associated with this, uncertainty over the delineation of sister-groups.

I will accept what appears to be the prevailing current view on the insect/myriapod relationship, namely: (a) that both of these groups are monophyletic; and (b) that they are sister-groups, within a broader group which may be termed Uniramia (Barnes, 1984) or Tracheata or Atelocera (Hennig, 1981). However, I must also note that several authors—notably Manton (1977)—disagree with the prevailing view.

Assuming that the Insecta (or what Kristensen (1981) calls Hexapoda: see Figure 5.3) is indeed the sister-group of the Myriapoda, the important question then becomes: at what stage do insects and myriapods begin to diverge in their development? It is difficult to give a simple answer to this question both because of the paucity of information on the development of myriapods and because what little information we do have shows that they, as well as the insects, are very embryologically diverse. That is, it is not possible to compare a single sort of development called 'insect development' with another sort called 'myriapod development': rather, the comparison is between two *groups* of ontogenies.

This problem can at least be lessened by concentrating on morphological features which are broadly common to each group but which differ between them; consequently I will focus, below, on the number and nature of segments. In insects, segmentation first occurs at a very early embryonic stage known as the germ-band. This stage follows the cellular blastoderm stage, and arises from it by cell proliferation at, and cell migration towards, a strip or band of tissue running in an anterior-posterior direction along the mid-ventral part of the blastoderm (see Figure 5.4 and Slack, 1983, Chapter 4). It is this 'germ-band' that becomes segmented and eventually develops into the larval (and later the adult) insect. The *number* of segments appropriate to the larval form—usually 14 post-cephalic segments but sometimes less—is established at this stage.

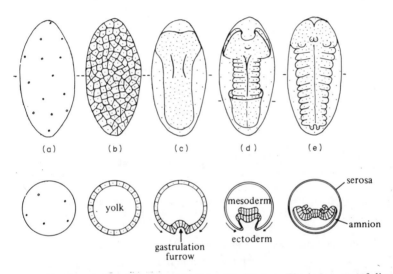

Figure 5.4 Embryonic development of a generalized insect. The lower row of diagrams shows transverse sections through the levels indicated on the upper row. (a) Cleavage, (b) blastoderm, (c) germ anlage, beginning of gastrulation (d) germ band, formation of embryonic membranes, (e) late germ band stage. (From Slack, 1983)

The contrast with the ontogeny of myriapod segment number is complicated by: (1) the variability in number within this group (though of course they 'typically' have more segments than do insects); and (2) the variability in the mode of production of segments, with some myriapods showing early establishment of the appropriate number, as insects do, and other showing gradual addition of segments at the posterior end throughout their development. Nevertheless, an insect/myriapod difference exists from the germ-band stage, though the nature of the difference depends on the group of myriapods that is chosen for comparison.

The centipede *Scolopendra cingulata* is an example of a myriapod in which

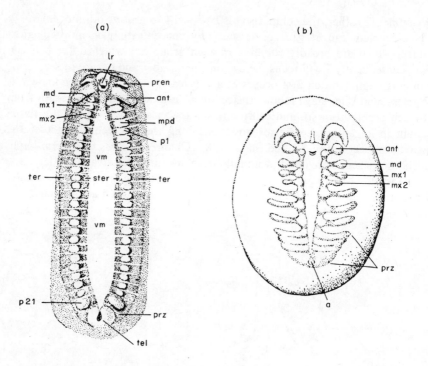

Figure 5.5 The germ-band stage of myriapods: (a) the centipede *Scolopendra cingulata*; (b) the millipede *Platyrhacus amauros*; lr, labrum; pren, preantenna; ant, antenna; md, mandible; mx, maxilla; mpd, maxilliped; p1-p21, postcephalic segments; prz, proliferation zone; tel, telson; a, anus; ter, tergite; ster, sternite; vm, membrane ventralis. (From Johannsen and Butt, 1941)

the adult segment number is laid down 'all at once' at the germ-band stage (see Figure 5.5(a)): 21 post-cephalic segments in this case. Here, we see an insect/myriapod developmental divergence at a very early stage of ontogeny. There is an obvious contrast with the Diptera/Mecoptera comparison, in which no such difference would be apparent.

In some other centipedes, and in millipedes, segmentation also begins at the germ-band stage, but the initial number of segments is low, and there is a gradual increase in the number of segments in 'later development'. How much later depends on the group concerned. For example, the germ-band stage of the millipede *Platyrhacus amauros* shown in Figure 5.5(b) has only three post-cephalic segments, whereas the adult millipede, of course, has very many more. So again we see a difference between insects and myriapods in segment number at an early stage, but the difference is of a more complex nature than before. In the language of morphogenetic tree theory, the comparison of the 'generalized insect' with *Platyrhacus* involves phase and/or structural changes being complicated by the superimposition of distortional changes.

Finally, I should stress that a comparison between insects and myriapods based on the *nature* of the segments—and in particular on the distinction between thoracic and abdominal ones—brings us to a similar conclusion to that reached on the basis of segment *number*, namely that there is early developmental divergence. Insects, which are generally characterized by a well-demarcated thorax and abdomen (an autapomorphy of the Insecta, to use the cladistic term), show this distinction from an early developmental stage, while myriapods never show such a distinction. Although the insect germ-band may not exhibit a difference between thoracic and abdominal segments (see Figure 5.4(e)), there is not long to wait for this difference to appear. For example in Diplura, which are probably closer than most extant insects to the stem form, the germ-band is first unsegmented, but shortly (a) becomes segmented and (b) shows a clear difference between the leg-bearing thoracic segments and the legless abdominal ones (Figure 5.6).

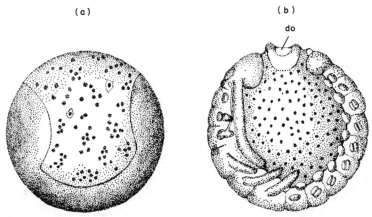

Figure 5.6 Early (a) and late (b) germ-band stages of the dipluran *Campodea staphylinus*. do, dorsal organ. (From Johannsen and Butt, 1941)

In the preceding case study it is clear not only that 'higher' taxa are characterized by more major differences in adult body plan than are 'lower' ones, but also that these differences arise earlier in development. I suspect that most readers would be prepared to accept this as the typical situation (though not, of course, the universal one) in the living world. Indeed, they might even have accepted it without my going into the details of a particular comparison. It may be that the point I am trying to make—that there is a link between the rank of taxonomic group and earliness of developmental divergence—is seen as 'obvious'. Yet I have taken some trouble to make this point (which incidentally is *not* just a restatement of von Baer's law since here taxonomic rank, as well as developmental time, is a variable) because I think that its implications for evolutionary theory are considerable and have not been fully appreciated.

The main point, to which I briefly alluded earlier, is as follows. If a morphological comparison is made between adults of two distantly related groups, large and numerous differences are observed. Given the usual situation of only fragmentary data on ancestral forms, a current large difference could have arisen either macromutationally or by a compounding of many individually negligible mutational changes. If we base the comparison on any single fixed point in the life-cycle other than the adult, the problem remains. But once we add in a developmental dimension, so that mutational changes are seen both in terms of magnitude *and* period of developmental effect (and indeed the interaction between the two), then many 'small' mutational changes (e.g. multiple occurrence of 'small triangles' in Figure 5.2) do *not* mimic a single 'large' one. They may do so in the magnitude dimension considered alone, but they do not do so in the time dimension.

To make an illustrative analogy: the time dimension is functioning in the same way as a max/min thermometer. In the case of the thermometer, several small dips in temperature from a prevailing level will be recorded differently to a single large drop. In the case of development, several accumulated small mutational changes will be 'recorded' (in the sense of information stored in a cross-taxon comparison) differently to a single, larger mutational change which penetrates much further back into development.

If there is a link of the kind proposed above between level of taxonomic group and time of developmental divergence, this clearly suggests that evolution operates by incorporating mutations of variable magnitude of effect, including (albeit rarely) some that would be classified as macromutations, according to our earlier working definition of that term. It argues against the rather homogeneous view of the more purist of the neo-Darwinians that all morphological change incorporated into evolution is of a minor, polygenic nature. If the extreme neo-Darwinian view is to continue to appear as a viable one, I think the onus is on its proponents to cast their view in a framework that includes developmental time *and* magnitude of effect, rather than just the latter, and to propose exactly how, in such a two-dimensional framework, they would picture a purely micromutational evolutionary mechanism.

I should make it clear that I am not proposing, here, that a higher taxon originates in some entirely novel way that is quite different from speciation. Rather, the suggestion being put forward is that most speciation events involve relatively little developmental change and have products that are normally regarded as congeneric with their parent species. Multiple occurrence of such a 'typical' speciation produces a highly speciose genus such as *Drosophila*. However, on rare occasions, a speciation occurs which involves earlier and more significant developmental changes. Given proliferation of the 'aberrant' product into many daughter-species through multiple occurrence of 'typical' speciations after the initial, unusual one, we are left with a group of species which are seen as variations on a common plan, a plan that is significantly different from that of the original parent-species. Each group characterized by a common plan is recognized in our taxonomy as a 'higher taxon', the precise level of which in the taxonomic hierarchy depends on, among other things, the earliness of developmental divergence initiated in the 'unusual' speciation. An added twist to this picture of evolution is that the early origins of high-level taxa (see Frazzetta, 1975) suggest that speciations involving radical developmental change were commoner in the early history of the biosphere—particularly in the Pre-Cambrian—than they are now. If so, there may have been both internal and external reasons for their comparative commonness in early evolution: less canalized developmental systems and more empty ecospace, respectively.

In Section 5.2, we saw how the available information on evolutionary rates (albeit very incomplete) is compatible with the prediction of morphogenetic tree theory that early developmental stages and the genes controlling them evolve more slowly than later ones. And the 'data' of the present section—namely the comparisons involved in the insect case-study—appear to support the associated prediction that evolution operates predominantly in a micromutational fashion, but that occasionally mutations of early-acting genes with larger effects on development must also make an input into the process, and that they will often be implicated in the origin of the body plans that characterize the higher taxa.

Of course, it would be nice to be able to go further than this rather general prediction about the origin of higher taxa and to bring data to bear on the relative frequency of:

1. *Advantageous* macromutations ('morphological windows').
2. Maladaptive ones (in the conventional sense) which become established through n-selection.
3. Conventional micromutational change coupled with increasing complexity of development (primary and secondary divergence).

At the present stage in our understanding of long-term evolution, this does not seem a realistic approach. What *does* seem perfectly realistic, however, is for many more case-studies to be conducted on comparison of the time of developmental divergence of taxa at different levels. Unfortunately, such comparative embryological studies are currently out of favour, but I would suggest that they still have much to tell us about evolution.

5.4 CONCLUSIONS

This section will serve a dual role: first as a conclusion for the chapter and second for the book as a whole. In both cases, the main conclusion can be stated very briefly.

On the application/testing of morphogenetic tree theory, with which this chapter has been concerned: the case-studies that have been discussed reveal an apparent compatibility of theory and (admittedly limited) data. However, there is the perennial problem that while incompatibilities between a theory and those parts of the real world which it attempts to describe are sufficient to refute, or at least to restrict the generality of, the theory, compatibility yields no clear conclusion. This is because there may be alternative theories that explain the data equally well or better. Also, we encounter a further problem here, namely that morphogenetic tree theory is potentially very broad in scope, giving a vast number of possible systems for its application/testing. Thus it is perhaps hardly surprising that its architect, who is naturally disposed towards supporting it, can come up with a set of compatible data. What will be much more telling is whether the theory's opponents can come up with data-sets with which it is *not* compatible. I look forward to seeing their attempts.

On a more general note: I hope that this book has succeeded in making a point which transcends morphogenetic tree theory, in the sense that it is true regardless of whether that theory does or does not ultimately gain acceptance. This point—which was first made in the preface and which I hope the intervening chapters have reinforced—is that a complete theory of morphological evolution must include a 'creative' component as well as a 'destructive' one. That is, it must include a treatment of the causal structure of development and its modification by mutation just as much as it must include a treatment of how the modified organisms fare in competition with each other. Even if the causal structure is not hierarchical, as suggested here, there must be *some* causal structure, and the form that it takes cannot be without relevance to evolution. And even if the distributions of 'magnitude of effect' suggested in the mutational version of the morphogenetic tree turn out to be the wrong ones, there must be *some* such distributions, and again their form must have evolutionary relevance. Such considerations suggest that, whatever the eventual fate of morphogenetic tree theory, the days of morphological evolutionary theories that treat development as a 'black box' are numbered.

References

Arthur, W. (1982a). Control of shell shape in *Lymnaea stagnalis*. *Heredity*, **49** 153–161.
Arthur, W. (1982b). A developmental approach to the problem of variation in evolutionary rates. *Biol. J. Linn. Soc.*, **18**, 243–261.
Arthur, W. (1984). *Mechanisms of Morphological Evolution: A Combined Genetic, Developmental and Ecological Approach.* Wiley, Chichester and New York.
Arthur, W. (1987a). *Theories of Life: Darwin, Mendel and Beyond.* Penguin, Harmondsworth, Middlesex.
Arthur, W. (1987b). *The Niche in Competition and Evolution.* Wiley, Chichester and New York.
von Baer, K.E. (1828). *Uber Entwicklungsgeschichte der Tiere: Beobachtung und Reflexion.* Borntrager, Konigsberg. (Part translation in Henfrey and Huxley, 1853.)
Balinsky, B.I. (1981). *An Introduction to Embryology*, 5th edn. Saunders College Publishing, Philadelphia.
Barnes, R.D. (1980). *Invertebrate Zoology*, 4th edn. Saunders College Publishing, Philadelphia.
Barnes, R.S.K. (1984). *A Synoptic Classification of Living Organisms.* Blackwell, Oxford.
de Beer, G. (1958). *Embryos and Ancestors*, 3rd edn. Clarendon Press, Oxford.
Bishop, J.A., and Cook, L.M. (1980). Industrial melanism and the urban environment. *Adv. Ecol. Res.*, **11**, 373–404.
Boncinelli, E., Simeone, A., La Volpe, A., Faiella, H., Acampora, D., and Scotto, L. (1985). Human cDNA clones containing homeobox sequences. *Cold Spring Harb. Symp. Quant. Biol.*, **50**, 301–306.
Bonner, J.T. (1974). *On Development: The Biology of Form.* Harvard University Press, Cambridge, Massachusetts.
Boudreaux, H.B. (1979). *Arthropod Phylogeny with Special Reference to Insects.* Wiley, New York.
Boycott, A.E. and Diver, C. (1923). On the inheritance of sinistrality in *Lymnaea peregra* (Mollusca, Pulmonata). *Proc. R. Soc. Lond. B*, **95**, 207–213.
Boycott, A.E., Diver, C., Garstang, S.L., and Turner, F.M. (1930). The inheritance of sinistrality in *Lymnaea peregra* (Mollusca, Pulmonata). *Phil. Trans. R. Soc. B*, **219**, 51–131.
Britten, R.J., and Davidson, E.H. (1969). Gene regulation for higher cells: a theory. *Science*, **165**, 349–357.
Carrasco, A.E., McGinnis, W., Gehring, W.J. and DeRobertis, E.M. (1984). Cloning of an *X. laevis* gene expressed during early embryogenesis coding for a peptide region homologous to *Drosophila* homeotic genes. *Cell*, **37**, 409–414.
Clarke, B. (1970). Darwinian evolution of proteins. *Science*, **168**, 1009–1011.

Clarke, B. and Murray J. (1969). Ecological genetics and speciation in land snails of the genus *Partula*. *Biol. J. Linn, Soc.*, **1**, 31–42.
Cohen, J. (1985). Review of *Mechanisms of Morphological Evolution*. *Biologist*, **32**, 54–55.
Cook, L.M. (1967). The genetics of *Cepaea nemoralis*. *Heredity*, **22**, 397–410.
Cook, L.M. (1971). *Coefficients of Natural Selection*. Hutchinson, London.
Cook, L.M., and Cain, A.J. (1980). Population dynamics, shell size and morph frequency in experimental populations of the snail *Cepaea nemoralis* (L.). *Biol. J. Linn. Soc.*, **14**, 259–292.
Coyne, J.A. (1983). Genetic basis of differences among three sibling species of *Drosophila*. *Evolution*, **37**, 1101–1118.
Crick, F.H.C. (1970). Diffusion in embryogenesis. *Nature*, **225**, 420–422.
Darwin, C. (1859). *On the Origin of Species by Means of Natural Selection, or the Preservation of Favoured Races in the Struggle for Existence*. John Murray, London.
Darwin, F. (ed.) (1887). *The Life and Letters of Charles Darwin*. John Murray, London.
Davidson, E.H., and Britten, R.J. (1979). Regulation of gene expression: possible role of repetitive sequences. *Science*, **204**, 1052–1059.
Degner, E. (1952). Der Erbang der Inversion bei *Laciniaria biplicata* MTG (Gastr. Pulm.). *Mitt. Hamburg zool. Mus.*, **51**, 3–61.
Desplan, C., Theis, J., and O'Farrell, P.H. (1985). The *Drosophila* developmental gene *engrailed* encodes a sequence-specific DNA binding activity. *Nature*, **318**, 630–635.
Eldredge, N., and Gould, S.J. (1972). Punctuated equilibria: an alternative to phyletic gradualism. In: *Models in Paleobiology*, Schopf, T.J.M. (ed.) Freeman, San Fransisco.
Erwin, D.H., and Valentine, J.W. (1984). 'Hopeful monsters', transposons, and Metazoan radiation. *Proc. Nat. Acad. Sci. USA*, **81**, 5482–5483.
Falconer, D.S. (1981). *Introduction to Quantitative Genetics*, 2nd edn. Longman, London.
Fincham, J.R.S. (1983). *Genetics*. John Wright, Bristol.
Fisher, R.A. (1930). *The Genetical Theory of Natural Selection*. Clarendon Press, Oxford.
Forey, P. (1985). Review of *Mechanisms of Morphological Evolution*. *Cladistics*, **1**, 396–399.
Frazzetta, T.H. (1975). *Complex Adaptations in Evolving Populations*. Sinauer, Sunderland, Massachusetts.
Freeman, G., and Lundelius, J.W. (1982). The developmental genetics of dextrality and sinistrality in the gastropod *Lymnaea peregra*. *Wilhelm Roux's Archives*, **191**, 69–83.
Garcia-Bellido, A. (1985). A gedanken exercise on mega-evolution. *Trends in Genetics*, **1**, 191.
Garrod, D.R. (1973). *Cellular Development*. Chapman & Hall, London.
Gaunt, S.J., Miller, J.R., Powell, D.J., and Dubole, D. (1986). Homeobox gene expression in mouse embryos varies with position by the primitive streak stage. *Nature*, **324**, 662–664.
Gehring, W.J., and Hiromi, Y. (1986). Homeotic genes and the homeobox. *Ann. Rev. Genet*, **20**, 147–173.
Gierer, A., and Meinhardt, H. (1972). A theory of biological pattern formation. *Kybernetic*, **12**, 30–39.
Goldschmidt, R. (1940). *The Material Basis of Evolution*. Yale University Press, New Haven, Connecticut.
Goldschmidt, R. (1952). Homeotic mutants and evolution. *Acta Biotheoretica*, **10**, 87–104.
Goodwin, B.C., and Cohen, M.H. (1969). A phase-shift model for the spatial and temporal organization of developing systems. *J. Theor. Biol.*, **25**, 49–107.

Gorczynski, R.M., and Steele, E.J. (1980). Inheritance of acquired immunological tolerance to foreign histocompatibility proteins in mice. *Proc. Nat. Acad. Sci. USA*, **77**, 2871–2875.

Gould, S.J. (1977a). Eternal metaphors of palaeontology. In: *Patterns of Evolution*, Hallam, A. (ed.) Elsevier, Amsterdam.

Gould, S.J. (1977b). *Ontogeny and Phylogeny*. Harvard University Press, Cambridge, Massachusetts.

Gould, S.J. (1982). The meaning of punctuated equilibrium and its role in validating a hierarchical approach to macroevolution. In: *Perspectives on Evolution*, Milkman, R. (ed.) Sinauer, Sunderland, Massachusetts.

Gould, S.J. (1985). The paradox of the first tier: an agenda for paleobiology. *Paleobiology*, **11**, 2–12.

Gould, S.J., and Eldredge, N. (1977). Punctuated equilibria: the tempo and mode of evolution reconsidered. *Paleobiology*, **3**, 115–151.

Gould, S.J., and Eldredge, N. (1986). Punctuated equilibrium at the third stage. *Syst. Zool.*, **35**, 143–148.

Gould, S.J., and Lewontin, R.C. (1979). The Spandrels of San Marco and the Panglossian paradigm: a critique of the adaptationist programme. *Proc. R. Soc. Lond. B*, **205**, 581–598.

Gould, S.J., Young, N.D., and Kasson, B. (1985). The consequences of being different: sinistral coiling in *Cerion*. *Evolution*, **39**, 1364–1379.

Haeckel, E. (1866). *Generelle Morphologie der Organismen*. Georg Reimer, Berlin.

Hamburger, V. (1980). Embryology and the modern synthesis in evolutionary theory. In: *The Evolutionary Synthesis: Perspectives on the Unification of Biology*, Mayr, E., and Provine, W.B. (ed.) Harvard University Press, Cambridge, Massachusetts.

Henfrey, A., and Huxley, T.H. (1853). *Scientific Memoirs, Selected from the Transactions of Foreign Academies of Science, and from Foreign Journals: Natural History*. Taylor & Francis, London.

Hennig, W. (1966). *Phylogenetic Systematics*. University of Illinois Press, Urbana.

Hennig, W. (1981). *Insect Phylogeny*. Wiley, Chichester and New York.

Holland, P.W.H., and Hogan, B.L.M. (1986). Phylogenetic distribution of *Antennapedia*-like homeoboxes. *Nature*, **321**, 251–253.

Ingham, P.W. and Martinez-Arias, A. (1986). The correct activation of *Antennapedia* and bithorax complex genes requires the *fushi tarazu* gene. *Nature*, **324**, 592–597.

Johannsen, O.A., and Butt, F.H. (1941). *Embryology of Insects and Myriapods*. McGraw-Hill, New York.

Jones, J.S. (1984). Developing theories of evolution. *Nature*, **312**, 386.

Jones, J.S., Leith, B.H., and Rawlings, P. (1977). Polymorphism in *Cepaea*: a problem with too many solutions? *Ann. Rev. Ecol. Syst.*, **8**, 109–143.

Kimura, M. (1968). Genetic variability maintained in a finite population due to production of neutral and nearly neutral isoalleles. *Genet. Res.*, **11**, 247–269.

Kimura, M. (1983). *The Neutral Theory of Molecular Evolution*. Cambridge University Press, Cambridge.

Koestler, A. (1970). *The Ghost in the Machine*. Pan, London.

Kollar. E.J., and Fisher, C. (1980). Tooth induction in chick epithelium: expression of quiescent genes for enamel synthesis. *Science*, **207**, 993–995.

Krebs, C.J. (1985). *Ecology: The Experimental Analysis of Distribution and Abundance*, 3rd edn. Harper & Row, New York.

Kristensen, N.P. (1981). Phylogeny of insect orders. *Ann. Rev. Entomol.*, **26**, 135–157.

Laughon, A., and Scott, M.P. (1984). Sequence of a *Drosophila* segmentation gene: protein structure homology with DNA- binding proteins. *Nature*, **310**, 25–31.

Levene, H. (1953). Genetic equilibrium when more than one ecological niche is available. *Am. Nat.*, **87**, 331–333.

Levine, M., Rubin, G.M., and Tjian, R. (1984). Human DNA sequences homologous to a protein coding region conserved between homeotic genes of *Drosophila*. *Cell*, **38**, 667–673.
Lewis, E.B. (1978). A gene controlling segmentation in *Drosophila*. *Nature*, **276**, 565–570.
Lewontin, R.C. (1974). *The Genetic Basis of Evolutionary Change*. Columbia University Press, New York.
Lewontin, R.C. (1985). Population genetics. *Ann. Rev. Genet.*, **19**, 81–102.
Lindsley, D.L., and Grell, E.H. (1968). Genetic Variations of *Drosophila melanogaster*. Carnegie Inst. Wash. Publ. 627.
Løvtrup, S. (1974). *Epigenetics: A Treatise on Theoretical Biology*. Wiley, London and New York.
McGinnis, W., Levine, M.S., Hafen, E., Kuroiwa, A., and Gehring, W.J. (1984a). A conserved DNA sequence in homeotic genes of the *Drosophila* Antennapedia and Bithorax complexes. *Nature*, **308**, 428–433.
McGinnis, W., Garber, R.L., Wirz, J., Kuroiwa, A., and Gehring, W. (1984b). A homologous protein-coding sequence in *Drosophila* homeotic genes and its conservation in other metazoans. *Cell*, **37**, 403–408.
McGinnis, W., Hart, C.P., Gehring, W.J., and Ruddle, F.H. (1984c). Molecular cloning and chromosome mapping of a mouse DNA sequence homologous to homeotic genes of *Drosophila*. *Cell*, **38**, 675–680.
McGinnis, W. (1985). Homeobox sequences of the antennapedia class are conserved only in higher animal genomes. *Cold Spring Harb. Symp. Quant. Biol.*, **50**, 263–270.
Mahowald, A.P., and Hardy, P.A. (1985). Genetics of *Drosophila* embryogenesis. *Ann. Rev. Genet.*, **19**, 149–177.
Manton, S.M. (1977). *The Arthropoda: Habits, Functional Morphology and Evolution*. Clarendon Press, Oxford.
Mather, K. (1941). Variation and selection of polygenic characters. *J. Genet.*, **41**, 159–193.
Mather, K. (1943a). Polygenic inheritance and natural selection. *Biol. Rev.*, **18**, 32–64.
Mather, K. (1943b). Polygenic balance in the canalization of development. *Nature*, **151**, 68–71.
Mather, K. (1943c). Polygenes in development. *Nature*, **151**, 560.
Maynard Smith, J. (1969). The status of neo-Darwinism. In: *Towards a Theoretical Biology. 2. Sketches*, Waddington, C.H. (ed.) Edinburgh University Press, Edinburgh.
Maynard Smith, J., and Sondhi, K.C. (1961). The arrangement of bristles in *Drosophila*. *J. Embryol. Exp. Morphol.*, **9**, 661–672.
Mayr, E. (1963). *Animal Species and Evolution*. Harvard University Press, Cambridge, Massachusetts.
Meinhardt, H., and Gierer, A. (1974). Applications of a theory of biological pattern formation. *J. Cell Sci.*, **15**, 321–346.
Muller, M.M., Carrasco, A.E., and DeRobertis, E.M. (1984). A homeobox- containing gene expressed during oogenesis in *Xenopus*. *Cell*, **39**, 157–162.
Murray, J. (1972). *Genetic Diversity and Natural Selection*. Oliver & Boyd, Edinburgh.
Nüsslein-Volhard, C., and Wieschaus, E. (1980). Mutations affecting segment number and polarity in *Drosophila*. *Nature*, **487**, 795–801.
Ohno, S. (1970). *Evolution by Gene Duplication*. Springer-Verlag, New York.
Patterson, C. (1982). Morphological characters and homology. In: *Problems of Phylogenetic Reconstruction*, Joysey, K.A., and Friday, A.E. (eds.) Academic Press, London.
Patterson, C., and Smith, A.B. (1987). Is the periodicity of extinctions a taxonomic artefact? *Nature*, **330**, 248–251.
Peel, R.A. (1986). Review of *Mechanisms of Morphological Evolution*. *Biology & Society*, **3**, 43–44.

Pelseneer, P. (1920). *Les Variations et leur Heredite Chez les Mollusques*. M. Hayez, Bruxelles.
Platnick, N.I. (1979). Philosophy and the transformation of cladistics. *Syst. Zool.*, **28**, 537–546.
Pritchard, D.J. (1986). *Foundations of Developmental Genetics*. Taylor & Francis, London and Philadelphia.
Raff, R.A., and Kaufman, T.C. (1983). *Embryos, Genes and Evolution: The Developmental Basis of Evolutionary Change*. Macmillan, New York.
Rasmussen, N. (1987). A new model of developmental constraints as applied to the *Drosophila* system. *J. Theor. Biol.*, **127**, 271–299.
Raup, D.M., and Sepkoski, J.J. (1984). Periodicity of extinctions in the geological past. *Proc. Nat. Acad. Sci. USA*, **81**, 801–805.
Riedl, R. (1978). *Order in Living Organisms: A Systems Analysis of Evolution*. Wiley, Chichester and New York.
Robertson, M. (1987). A genetic switch in *Drosophila* morphogenesis. *Nature*, **327**, 556–557.
Rosen, D. (1984). Hierarchies and history. In: *Evolutionary Theory: Paths into the Future*, Pollard, J.W. (ed.) Wiley, Chichester and New York.
Sang, J. (1984). *Genetics and Development*. Longman, London and New York.
Schank, J.C., and Wimsatt, W.C. (1986). Generative entrenchment and evolution. In: *PSA-1986*, volume 2, eds. P.K. Machamer & A.T. Fine. Philosophy of Science Association, East Lansing, Michigan.
Schmalhausen, I.I. (1949). *Factors of Evolution*. Blakiston, Philadelphia.
Scott, M.P., and Weiner, A.J. (1984). Structural relationships among genes that control development: sequence homology between the Antennipedia, Ultrabithorax, and fushi tarazu loci of *Drosophila*. *Proc. Nat. Acad. Sci. USA*, **81**, 4115–4119.
Shepherd, J.C.W., McGinnis, W., Carrasco, A.E., DeRobertis, E.M., and Gehring, W.J. (1984). Fly and frog homeo domains show homologies with yeast mating type regulatory proteins. *Nature*, **310**, 70–71.
Shumway, W. (1932). The recapitulation theory. *Q. Rev. Biol.*, **7**, 93–99.
Siegel, S. (1956). *Non-parametric Statistics for the Behavioral Sciences*. McGraw-Hill Kogakusha, Tokyo.
Simpson, G.G. (1944). *Tempo and Mode in Evolution*. Columbia University Press, New York.
Simpson, G.G. (1953). *The Major Features of Evolution*. Columbia University Press, New York.
Sinnott, E.W., Dunn, L.C., and Dobzhansky, T. (1958). *Principles of Genetics*, 5th edn. McGraw-Hill Kogakusha, Tokyo.
Slack, J.M.W. (1983). *From Egg to Embryo: Determinative Events in Early Development*. Cambridge University Press, Cambridge.
Stanley, S.M. (1975). A theory of evolution above the species level. *Proc. Nat. Acad. Sci. USA*, **72**, 646–650.
Stanley, S.M. (1979). *Macroevolution: Pattern and Process*. Freeman, San Fransisco.
Struhl, G. (1981). A gene product required for correct initiation of segmental determination of *Drosophila*. *Nature*, **293**, 36–41.
Sturtevant, A.H. (1923). Inheritance of direction of coiling in *Limnaea*. *Science*, **58**, 269–270.
Thom, R. (1983). *Mathematical Models of Morphogenesis*. Ellis Horwood, Chichester.
Turing, A.M. (1952). The chemical basis of morphogenesis. *Phil. Trans. R. Soc. B*, **237**, 37–72.
de Vries, H. (1905). *Species and Varieties, Their Origin by Mutation*. Open Court, Chicago.
Waddington, C.H. (1940). *Organizers and Genes*. Cambridge University Press, Cambridge.

Waddington, C.H. (1942). Canalization of development and the inheritance of acquired characters. *Nature*, **150**, 563–565.
Waddington, C.H. (1943). Polygenes and oligogenes. *Nature*, **151**, 394.
Waddington, C.H. (1953). The genetic assimilation of an acquired character. *Evolution*, **7**, 118–126.
Waddington, C.H. (1956). Genetic assimilation of the bithorax phenotype. *Evolution*, **10**, 1–13.
Waddington, C.H. (1957). *The Strategy of the Genes*. Allen & Unwin, London.
Waddington, C.H. (1975). *The Evolution of an Evolutionist*. Edinburgh University Press, Edinburgh.
Wieschaus, E., and Gehring, W. (1976). Clonal analysis of primordial disc cells in the early embryo of *Drosophila melanogaster*. *Dev. Biol.*, **50**, 249–263.
Williamson, P.G. (1981). Morphological stasis and developmental constraint: real problems for neo-Darwinism. *Nature*, **294**, 214–215.
Willis, J.C. (1940). *The Course of Evolution by Differentiation or Divergent Mutation rather than by Selection*. Cambridge University Press, Cambridge.
Wimsatt, W.C. (1986). Developmental constraints, generative entrenchment, and the innate-acquired distinction. In: *Integrating Scientific Disciplines*, Bechtel, W. (ed.) Martinus- Nijhoff, Dordrecht.
Wolpert, L. (1968). The French flag problem: a contribution to the discussion on pattern formation and regulation. In: *Towards a Theoretical Biology. 1. Prolegomena*, Waddington, C.H. (ed.) Edinburgh University Press, Edinburgh.
Wolpert, L. (1969). Positional information and the spatial pattern of cellular differentiation. *J. Theor. Biol.*, **25**, 1–47.
Wolpert, L. (1971). Positional information and pattern formation. *Curr. Top. Dev. Biol.*, **6**, 183–224.

Author Index

Acampora, D. 71
Arthur, W. 2, 5, 8, 11, 12, 14, 20, 21, 24, 26, 29, 31, 37, 39, 42, 44, 45, 50, 51, 59, 63, 69, 72

von Baer, K.E. 2, 22
Balinsky, B.I. 28
Barnes, R.D. 38
Barnes, R.S.K. 76, 78
de Beer, G. 24, 28, 30, 32, 48
Bishop, J.A. 37
Boncinelli, E. 71
Bonner, J.T. 28, 67
Boudreaux, H.B. 76
Boycott, A.E. 18, 43
Britten, R.J. 8, 19
Butt, F.H. 80, 81

Cain, A.J. 61
Carrasco, A.E. 70–72
Clarke, B. 17, 43
Cohen, J. 50
Cohen, M.H. 3
Cook, L.M. 37, 45, 61
Coyne, J.A. 76, 78
Crick, F.H.C. 4

Darwin, C. 1, 18, 36, 56
Darwin, F. 36
Davidson, E.H. 8, 19
Degner, E. 43
DeRobertis, E.M. 70–72
Desplan, C. 70
Diver, C. 18, 43
Dobzhansky, T. 46
Dubole, D. 71
Dunn, L.C. 46

Eldredge, N. 37, 59, 60
Erwin, D.H. 42, 46

Faiella, H. 71
Falconer, D.S. 35
Fincham, J.R.S. 17, 18
Fisher, C. 22
Fisher, R.A. 24, 26
Forey, P. 50
Frazzetta, T.H. 83
Freeman, G. 18, 21

Garber, R.L. 70
Garcia-Bellido, A. 29
Garrod, D.R. 12
Garstang, S.L. 18, 43
Gaunt, S.J. 71
Gehring, W.J. 70–72, 74
Gierer, A. 3
Goldschmidt, R. 1, 48, 53–56
Goodwin, B.C. 3
Gorczynski, R.M. 37
Gould, S.J. 1, 2, 21, 22, 28, 30, 32, 37, 40, 43, 55, 59, 60, 62, 67
Grell, E.H. 31

Haeckel, E. 22, 30
Hafen, E. 70
Hamburger, V. 57
Hardy, P.A. 35
Hart, C.P. 71
Hennig, W. 61, 76, 78
Hiromi, Y. 70
Hogan, B.L.M. 70
Holland, P.W.H. 70

Ingham, P.W. 14

Johannsen, O.A. 80, 81
Jones, J.S. 43, 50, 72

Kasson, B. 21, 43
Kaufman, T.C. 68

Kimura, M. 2, 24, 37, 51
Koestler, A. 8
Kollar, E.J. 22
Krebs, C.J. 45
Kristensen, N.P. 76–78
Kuroiwa, A. 70

La Volpe, A. 71
Laughon, A. 70
Leith, B.H. 43
Levene, H. 47
Levine, M. 70, 72
Lewis, E.B. 69
Lewontin, R.C. 1, 21, 51, 55
Lindsley, D.L. 31
Løvtrup, S. 9, 56
Lundelius, J.W. 18, 21

McGinnis, W. 70–72
Mahowald, A.P. 35
Manton, S.M. 78
Martinez-Arias, A. 14
Mather, K. 17, 29, 59
Maynard Smith, J. 11, 36
Mayr, E. 36, 53
Meinhardt, H. 3
Miller, J.R. 71
Muller, M.M. 71
Murray, J. 43, 47

Nüsslein-Volhard, C. 45, 69

O'Farrell, P.H. 70
Ohno, S. 30, 41

Patterson, C. 28, 62, 63, 67
Peel, R.A. 50
Pelseneer, P. 43
Platnick, N.I. 61
Powell, D.J. 71
Pritchard, D.J. 31

Raff, R.A. 68
Rasmussen, N. 75
Raup, D.M. 62
Rawlings, P. 43

Riedl, R. 9, 54, 56
Robertson, M. 70
Rosen, D. 62
Rubin, G.M. 72
Ruddle, F.H. 71

Sang, J.H. 8
Schank, J.C. 13
Schmalhausen, I.I. 56
Scott, M.P. 70
Scotto, L. 71
Sepkoski, J.J. 62
Shepherd, J.C.W. 70
Shumway, W. 6
Siegel, S. 62
Simeone, A. 71
Simpson, G.G. 37, 44
Sinnott, E.W. 46
Slack, J.M.W. 3, 8, 12, 39, 79
Smith, A.B. 63
Sondhi, K.C. 11
Stanley, S.M. 37, 44, 60
Steele, E.J. 37
Struhl, G. 18
Sturtevant, A.H. 18

Theis, J. 70
Thom, R. 12
Tjian, R. 72
Turing, A.M. 11
Turner, F.M. 18, 43

Valentine, J.W. 42, 46
de Vries, H. 53

Waddington, C.H. 8, 12, 13, 18, 36, 56–60
Weiner, A.J. 70
Wieschaus, E. 45, 69, 74
Williamson, P.G. 59–61
Willis, J.C. 53
Wimsatt, W.C. 13, 27
Wirtz, J. 70
Wolpert, L. 3, 4, 8, 12

Young, N.D. 21, 43

Subject Index

adaptationist programme 21
allele substitution 24–26, 67
allometry 6, 36

behaviour 2
body plan 34, 35, 65, 66, 73
branching track system 18, 57, 58
burden 27

caenogenesis 29
canalization 12, 13, 18, 56–60, 83
causal link 3, 4
cell-autonomous effect 3, 74, 75
cell differentiation 2, 12
chromosomal repatterning 54, 58
cladistics 61, 62
coadaptation 27, 31, 40, 42, 46, 47
competition 44, 46, 47, 84
complexity 22, 29–32
co-ordination 12
creode 57
cross-linking of processes 13, 57, 60

D-loci 14, 24, 69–73
DNA 69, 70
developmental constraint 1, 13, 21, 32, 40, 60, 68
developmental decisions 8, 16–18, 34
developmental heterogeneity 4, 11
developmental lock 13
distortional change 21, 22, 32
drift, genetic 24

embryo 6, 68
  mosaic 69
  regulatory 69
embryonic field 8
enzyme 2
epigenetic landscape 8, 18, 57, 58
extinction 62, 63

fitness 29, 45
French flag model 4, 5

gamete 46
gene
  duplication 30, 70
  housekeeping 24
  major 27
  modifier 27, 37, 60
generative entrenchment 13, 27
genetic assimilation 18, 56
genome size 30
geographical race 61
gradient 3–6

heterochrony 32
hierarchy, developmental 6–13, 19, 56–58, 84
homeobox 70–73
homeorhesis 57, 60
homeostasis 60
homology 28, 61, 62
hopeful monster 48, 54, 58

imaginal disc 10, 78
independence of characters 26, 41, 42
induction 3, 4, 22
industrial melanism 37

Lamarckian inheritance 37
life cycle 5, 7, 10, 11, 29, 68, 82
life table 45

maternal inheritance 18, 69
modern synthesis 1, 2, 36, 56
molecular evolution 2, 51
morphogen 3, 4
morphogenetic tree theory, summary 51–53
morphogenesis 12

mutation
  macro- 35–37, 53–56, 73, 74
  magnitude of effect 15–18, 26, 44, 56, 64, 73, 74, 84
  micro- 17, 35–38, 53–56, 73
  rate 24, 26, 31
  systemic 54

n-selection 45, 46, 54, 63, 83
natural classification 27, 61, 62
neo-Darwinism 17, 20, 28, 35–37, 43, 44, 51, 53–56, 62, 63, 82
neoteny 32, 40
neutral allele 24
niche 44, 47, 54

oligogene 17, 29

pattern formation 2, 12, 19
phase change 21, 24–29
phenotypic evolution 27, 67
pigmentation 42, 43
polygene 17, 29, 36, 59, 71

polymorphism 47, 51
positional information 12
preadaptation 54
progenesis 32, 40
punctuated equilibrium 37, 59–61

recapitulation 30, 39
regional specification 12, 19
repeatability 12

segmentation 69–83
selection
  natural 1, 37
  species 59–61
  stabilizing 60
speciation 60, 82
stasis, morphological 59–61
structural change 21, 22, 29–32
survivorship curve 25

virus 46
von Baer's law 24–29, 30, 62, 67–69, 81